Proceedings of the Seventh Annual Conference of the British Association for Biological Anthropology and Osteoarchaeology

Edited by

Sonia R. Zakrzewski
William White

BAR International Series 1712

2007

Published in 2016 by
BAR Publishing, Oxford

BAR International Series 1712

Proceedings of the Seventh Annual Conference of the British Association for Biological Anthropology and Osteoarchaeology

ISBN 978 1 4073 0156 3

BAR Publishing is the trading name of British Archaeological Reports (Oxford) Ltd.
British Archaeological Reports was first incorporated in 1974 to publish the BAR
Series, International and British. In 1992 Hadrian Books Ltd became part of the BAR
group. This volume was originally published by Archaeopress in conjunction with
British Archaeological Reports (Oxford) Ltd / Hadrian Books Ltd, the Series principal
publisher, in 2007. This present volume is published by BAR Publishing, 2016.

Printed in England

BAR
PUBLISHING

BAR titles are available from:

BAR Publishing
122 Banbury Rd, Oxford, OX2 7BP, UK
EMAIL info@barpublishing.com
PHONE +44 (0)1865 310431
FAX +44 (0)1865 316916
www.barpublishing.com

Proceedings of the Seventh Annual Conference of the
British Association for Biological Anthropology and Osteoarchaeology

Contents

Introduction

William White & Sonia R. Zakrzewski

Centre for Human Bioarchaeology
Museum of London
150 London Wall
London, EC2Y 5HN

Dept of Archaeology
University of Southampton
Highfield
Southampton, SO17 1BF

This volume represents the third publication of the conference proceedings of the British Association for Biological Anthropology and Osteoarchaeology and consists of a sample of the papers and posters presented at the Association's Seventh annual conference Since its inception the Association has striven to include all aspects of research within Osteoarchaeology and the wider field of Biological Anthropology. The first conference was held in Birmingham, in 1999, and included sessions entitled "Planning, Excavation, Processing and Curation of Human Skeletal Remains" and "Human Growth and Reproduction". More recently, the conferences have included sessions which concentrated upon aspects of palaeopathology and comparative primate morphology and behaviour. These conferences have been notable for demonstrating the breadth and diversity in research within the field of biological anthropology, both within the UK and from further afield. The conferences have been highly successful at developing osteoarchaeology within Britain, through sharing the research being undertaken with a wider audience, and from developing British standards for the recording of human skeletal remains (Brickley & McKinley 2004).

The seventh conference was held at the Museum of London in 2005 and was attended by over 120 people. The sessions included in the conference ranged from "Commercial Archaeology in the Capital", through an open session in memory of Trevor Anderson to period-based sessions (e.g. prehistory and post-medieval). Many of the podium sessions included linked poster presentations, in addition to a dedicated poster session. This publication therefore presents a representative sample of the material presented within the general themes of the conference sessions.

All the papers submitted were subjected to full peer-review. As editors, we have been strict in following the requirements and comments made by the reviewers, and hence we feel that the following papers are all of a high standard. It is noteworthy that the papers themselves, as for the previous Proceedings, are written by a wide variety of researchers, from postgraduate students to professors, and also include some from researchers outside academia. We would like to thank the following (in alphabetical order) for their help, advice and dedicated reviewing of the papers submitted: Jelena Bekvalac, Megan Brickley, Anwen Caffell, Andrew Chamberlain, Lynne Cowal, Jenny Hall, Sarah Inskip, Tina Jakob, Tania Kausmally, Chris Knüsel, Mary Lewis, Simon Mays, Richard Mikulski, Piers Mitchell, John Robb, Charlotte Roberts, Louise Scheuer, Holger Schutkowski, and Martin Smith. We are very grateful to them for all the advice and support received. The volume is organised into three loose themes.

The first theme, comprising the first five papers within this volume and reflecting one of the major issues raised during the meeting, concerns the study of human remains from within London. This section determinedly includes research papers written by colleagues working in the more commercially-based sectors of archaeology and anthropology. It is important that the research potential of the material currently being excavated and curated within London is recognised both within and without the archaeological field. The first paper, by White, describes the history and development of human osteology within the Museum of London, and provides a description of the research potential of both the Wellcome Osteological Research Database (WORD) and the skeletal collections themselves curated within the Museum of London. Based upon experiences at St Pancras burial ground, the next paper, by Emery, summarises the issues arising from large scale construction works within London, and discusses the potential of on-site research within the contract-based situation. This is of particular importance given the varying legal situations that arise from such large scale works. The next two papers, by Melikian & Sayer and Bello & Humphrey, provide London-based case studies of human skeletal assemblages.

The second theme, comprising the next five papers, provides a series of individual case studies dealing, primarily, with palaeopathological topics. Syphilis within London is considered and assessed by both Powers & Emery and by Patel & Mitchell, but employing very different data sets. Ponce, Arabaolaza & Boylston summarise an example of potential amputation from Wolverhampton, whilst Zakrzewski & Morris describe a Caribbean specimen either exhibiting severe neck trauma, block vertebrae formation or Klippel-Feil syndrome. The last paper in this section, by Mikulski, continues the global perspective and describes the mortuary treatment and palaeopathology from a site in Peru.

The final section of the volume contains papers that consider and develop methods of analysis. Patrick evaluates the usefulness of skeletal material for estimating body mass and the potentially overweight and obese. Clement describes and evaluates a novel method for assessing dental wear, whereas Rose describes the employment of stable isotopes to access data on diet within in a North American context. Storm evaluates analysis of skeletal asymmetry and changes in the degree of this asymmetry occurring across time, and the last paper, by Buckberry & O'Connor evaluate the potential of radiography within a palaeopathological context.

We hope that these papers will demonstrate the diversity and depth of research in biological anthropology presented at the conference held at the Museum of London. Many other papers and posters were presented, and so the papers included here should be viewed as simply a selection of those presented. We would like to thank all those who presented their research at the conference, be it in podium presentation or poster format. We also would like to reiterate our thanks to all the reviewers for their care and attention. Thanks are due to Sarah Inskip for her help in the final editing of this volume. Lastly, but by no means least, we thank all the authors for their support and understanding. We hope that this volume will be a testament to their research.

Literature Cited

Brickley M and McKinley J (2004) Guidelines to the Standards for Recording Human Remains. IFA Paper Number 7. IFA/BABAO Publication. http://www.babao.org.uk/

An overview of human osteology at the Museum of London

William White

Centre for Human Bioarchaeology
Museum of London
150 London Wall
London
EC2Y 5HN
bwhite@museumoflondon.org.uk

Abstract

The Museum of London was established in 1974 by the merger of the collections of the London Museum, Kensington Palace, and the Guildhall Museum, City of London. From its inception the Museum had an archaeological arm involved in rescue excavation in the City. Over 30 years the Museum's Departments of Urban Archaeology, Greater London Archaeology and latterly the Museum of London Archaeology Service (MoLAS) have excavated thousands of archaeological sites, several hundred of which have produced human remains. The curated human skeletal remains from these excavations cover all historical periods and now account for more than 17,000 individuals. This is one of the biggest collections of stratified human remains in the world and is already the largest scientifically excavated and documented group from a single city. The collection attracts national and international interest. It represents a unique teaching collection and a resource that can and does support an extraordinary range of studies on, for example, palaeopathology, palaeodemography, population-based approaches to data analysis, ancient biomolecular research, bone chemical analysis and forensic anthropology. This is reflected in the fact that the Museum of London has now reached a full-time staff of 10 osteoarchaeologists recording skeletons directly onto an electronic database.

Keywords: Archaeology, burials, database, London

Since the summer of 2003, bioarchaeology has assumed a very high profile at the Museum of London. Research on London skeletons both internally (by the 10 full-time osteoarchaeologists on the Museum's staff) and externally (by postgraduate and post-doctoral scholars from a range of universities) is at a record high level. What follows here is concerned chiefly with the present level of activity and looking toward the future, but it is illuminating to look back at how the current situation has developed.

The Museum of London, following its foundation in 1974, immediately began to wield an archaeological arm, excavating sites in and around the City of London. Some of the excavations involved ancient cemeteries. Figure 1 summarises the cemetery sites, or other sites producing human remains, excavated during the first two decades of the Museum's existence. The late 1990s and the excavation in Greater London around the turn of the 21st century have been deliberately excluded because these dates are dominated by a single archaeological site, the Priory and Hospital of St Mary Spital, which changed the entire picture of human skeletal remains in London (Thomas 2002: 101-2; Thomas 2004). A peak of activity is visible for the late 1980s. This corresponds to an event known at the time as the "Big Bang", when deregulation of the financial markets in London led to a boom in office building and other building redevelopment in the City. The Museum of London Departments of Urban Archaeology and Greater London Archaeology responded to the considerable challenge of the rescue archaeology required and this was duly continued by

their successor, the Museum of London Archaeology Service (MoLAS).

As so often, post-excavation research on the cemetery sites lagged well behind excavation. This was partly because, although the Museum had a team of archaeozoologists working on faunal remains from its archaeological sites, it did not have a human bone specialist on its permanent staff. Osteological analysis was farmed out to freelance osteoarchaeologists around the country, including the author (White 1988). The Museum's first incumbent in the permanent post of human osteologist was Fiona Keily, in 1988, and she was succeeded in 1992 by Janice Conheeney. It was the latter who was chiefly responsible for the human bone reports in the stream of publications from MoLAS that have since appeared (Barber & Bowsher 2000; Barber et al. 2004; Bowsher et al. 2007; Cowan 2003; Drummond-Murray et al. 2002; Hiller & Wilkinson, 2005; Mackinder 2000; Malcolm et al. 2003; Miles, et al. 2007; Miller & Saxby 2007; Schofield & Lea 2005; Sidell et al. 2002; Sloane & Malcolm 2004; Swift 2003; Thomas et al. 1997; Watson 2003). As time went on, with the increase in archaeological skeletal remains from the London area to more than 17,000 individuals (one of the largest such collections in the world), the osteological resources needed to be increased. The Museum of London duly arrived at the current position, with four osteologists analysing and recording 5,600 skeletons from the site of the Priory and Hospital of St Mary Spital and a changing team of osteologists dealing with another c5,000 skeletons from London archaeological sites dating from

No of Sites

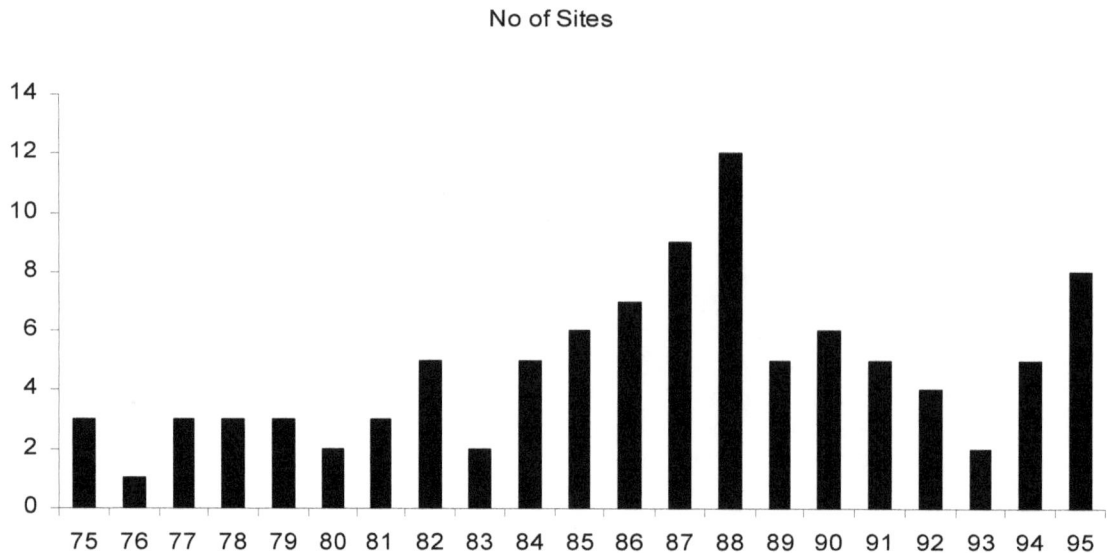

Figure 1. Total numbers of London archaeological sites yielding human remains by year of excavation (1900s)

the 1st to the early 19th centuries AD (there are no prehistoric skeletons, other than Bronze- and Iron Age cremation burials and these were not included in the analysis). The Project Manager was Gustav Milne, now of the Institute of Archaeology at UCL. In addition to this enormous effort, split between the two major projects, Natasha Powers had responsibility for contract work (including recent MoLAS sites with excavated skeletons) and also is the first point of contact in any forensic investigations in the City/Metropolitan Police districts. She is now Head of Osteology at MoLAS.

Although the members of the osteology teams may be working on different projects, what they have in common is that they record skeletons to a common standard and onto the same electronic database. The background to this is that the MoLAS paper skeletal recording sheets had grown haphazardly rather than rationally and eventually comprised a 20-page document for each skeleton. Brian Connell re-designed the recording sheets into a more compact and rational form. This design lent itself better to conversion into an electronic database and this was accomplished with Brian working in close association with Peter Rauxloh and the Museum of London's recording sheets. The fields in the database correspond closely to the pages of the paper recording forms and, for the first time, allow rapid direct data capture, without the need for an intermediate paper stage – something our archaeozoologist colleagues at the Museum had enjoyed for several years in recording archaeological faunal data. Paper recording sheets are now a rarity, except for certain fieldwork.

Making entries onto the database

In the first field, the Level 1 catalogue allows an inventory of all the bones present in the excavated skeleton to be compiled, using binary notation (presence/absence of bone regions), noting the condition of preservation and completeness of the bones recorded.

Then, at Level 2, subsequent fields allow for the age and sex assessment (and justifications) to be recorded, followed by the recording of metric and non-metric information to minimum standards. Palaeopathology recording makes use of free text on the database, and, overall, is based on the accurate description of the type of bone alteration and the distribution of lesions in the skeleton. This is as recommended by the British Association for Biological Anthropology and Osteoarchaeology (Roberts & Connell 2004). The testing and trouble-shooting phase of the database development began with the recording of the largest sample of human remains – that from the medieval Priory and Hospital of St Mary Spital.

Another initiative by Brian Connell was to publicise the Museum's larger samples of human remains and this led ultimately to the Museum's application to the Wellcome Trust for funding to analyse and record a diachronic collection of about 5,000 skeletons from the larger sites to a common standard, integrate archaeological site summaries and interpretative reports into an Oracle relational database and make the accumulated data accessible remotely via the internet. The Wellcome Osteological Research Database (WORD) encompasses 55 archaeological sites from London. Samples of large size were selected so that the data generated could be employed in population-based studies. The total numbers for each historical period are tabulated below (Table 1).

Table 1. Periods represented in Wellcome Osteological Research Database Project

Date	Number of individuals
Romano-British	975
Anglo-Saxon	80
Medieval [1]	2240
Post-medieval	1752
Total	5047

2

Major archaeological sites included in WORD

Roman Cemeteries in London were formally organised after the City wall was erected at the end of the second century AD. The pattern appeared to be of three major extramural cemeteries, at the ordinal points of the compass (west, north and east), with further cemeteries in Southwark (south of the River Thames (Hall 1986)). The Northern cemetery is dispersed either side of Bishopsgate (the Roman road running north out of the city of Londinium (Ermine Street)). The approximately 160 Romano-British burials from the Spitalfields site are not included in the total (Table 2), nor are the few earlier Roman intramural burials or the several hundred cremation burials curated in the Museum of London. Currently the Eastern Cemetery is the best-known, with 587 inhumations analysed in the definitive publication (Barber & Bowsher 2000). The southern and western cemeteries are less well known and only part-published (Bentley & Pritchard 1982; Mackinder 2000; Watson 2003).

Table 2. Inhumations in Romano-British cemeteries excavated in London (MoLAS)

Cemetery[1]	No. inhumations[2]
Western	65
Northern (Bishopsgate)[3]	80
Eastern[4]	600
Southern (i.e.Southwark)	84

Notes: [1] Dates AD 1st to 5th centuries, [2] Excluding cremation burials, [3] Excluding St Mary Spital Roman burials and certain other sites, [4] Published (Barber & Bowsher 2000)

In London, there is a paucity of known Anglo-Saxon burials. This is partly because the Roman city of Londinium was not re-populated on a significant scale until the 9th century AD under Alfred, who also re-fortified the City. A few chiefly late Saxon burials have been excavated within the City. The main Anglo-Saxon settlement was to the west of the City in the Aldwych/Covent Garden area in Westminster, where there are some scattered burials of 7th century date, but the main cemetery or cemeteries are yet to be located. Very recently, early Saxon burials have been identified at St Martin-in-the-Fields church in Trafalgar Square (Powers pers comm.).

The London Medieval samples are chiefly monastic in character (Gilchrist & Sloane 2005) and it would be helpful to have a few more large parochial samples (Table 3). The monastic samples are rather variable, depending upon the extent of archaeological excavation of the site concerned. Thus, at Bermondsey Abbey, the 200 individuals disinterred from the Cluniac monks' cemetery were male or of unknown sex; none were shown to be female and there were no juveniles. By contrast, excavations of Augustinian establishments

revealed different areas of burial (including inside the churches) with concentrations of monastic or lay burials (Miller & Saxby 2007; Thomas et al. 1997). The Royal Mint site has been listed in Table 3 as St Mary Graces but the Cistercian Abbey subsumed a slightly earlier burial ground for the mass burial of the victims of the Black Death epidemic of 1348-50, and these latter form the bulk of the sample. Understandably, this large sample has been in great demand for research of a diverse character (e.g, Antoine and Hillson 2005; Gilbert et al. 2004; Gowland & Chamberlain 2005; Margerison & Knüsel 2002; Rock et al. 2006; Waldron 1992, 2001; Wood & Dewitte 2003).

Table 3. Medieval burials in London: monastic and parochial (1066-1540)

Cemetery	No. burials
St Mary of Graces (Cistercian)	1030[1]
St Mary Merton, Surrey (Augustinian)[2]	679
St Saviour, Bermondsey (Cluniac)	200
Blackfriars, London (Dominican)	58
St John, Clerkenwell (Hospitaller)[3]	13[4]
Holy Trinity Priory[5]	105
St Nicholas Shambles[6]	234
St Lawrence Jewry	71
St Benet Sherehog	35[2]

Notes: [1] Includes 634 Black Death victims, [2] Published (Miller & Saxby 2007), [3] Published (Sloane & Malcolm 2004), [4] Sample too small to be included in the database, [5] Published (Schofield & Lea 2005), [6] Published (White 1988)

Post-medieval crypts, cemeteries and large samples have been coming to the fore in the wake of the flagship research sites; St Bride's Fleet Street (Milne 1997) and Christ Church Spitalfields (Molleson & Cox 1993), which, incidentally, are also London sites. As set out in Table 4, there is a broad gradation in socio-economic status from Chelsea at the upper end (Cowie et al. 2007) through to the paupers' burial ground in Redcross Way, Southwark (Brickley & Miles 1999). These sites generally cover the 18th and 19th centuries, but some are slightly earlier.

The raison d'être for the creation of the database was to inform and attract external researchers. Meanwhile, since 1998, a total of 90 postgraduate and post-doctoral scholars, from 30 academic institutions worldwide, have visited the Museum of London in order to conduct research on our holdings of human skeletal remains. There were a mere handful of applications for research in the early days but during the time of the immediate WORD pre-launch period they doubled and re-doubled to 24 successful applications per year. The Museum of London now experiences record numbers of applications from researchers to study the collections, thanks to the launch of the database online in April 2007.

Table 4. Table of post-medieval cemeteries excavated by the Museum of London

Cemetery	No. burials	Date	Social Status
Old Church Street, Chelsea[1]	197	1695-1846	High
St Benet Sherehog Burial Ground (No 1 Poultry)	230	c1700-1842	Middle
St Botolph Billingsgate	50	1595-1666	Middle
St Bride's Lower Churchyard, Farringdon Street	557	18th-19th C	Poor
St Thomas Hospital, New London Bridge	227	1540-1714	Poor
Broadgate, the 'New Churchyard'	388	1569-1720	Poor
Cross Bones Burial Ground (Redcross Way)[2]	166	18th-19th C	Pauper

Notes: [1] published (Cowie et al. 2007), [2] Published (Brickley et al. 1999)

Appendix: Data Downloads

The Wellcome Osteological Research Database is available as a microsite on the Museum of London's website, currently under development. Downloads are available as files that can be read into spreadsheet packages. For each cemetery the following elements are available: condition of bone preservation, average integrity of skeletal completeness, age estimates for individuals, sex estimates, dental anomalies, dental pathology, general pathology, joint surface assessment, vertebral degenerative joint disease, vertebral genetic anomalies, metadata summaries, etc.

Details are available on the website:

www.museumoflondon.org.uk/English/Collections/Online Resources/CHB/Resources/

In addition, more than 5,500 skeletons from the Priory and Hospital of St Mary Spital have already been recorded onto the database and the data will be rolled out online during 2009. Other sites will be added in due course.

Acknowledgements

The Wellcome Trust for Project Grant GR 7479AIA. The Spitalfields Development Group, Dr Peter Rauxloh and the Museum of London IT team. Archaeologists: Gustav Milne and Vanessa Bunton and Museum of London osteologists past and present: Jelena Bekvalac, Brian Connell, Lynne Cowal, Amy Gray-Jones,Tania Kausmally, Rebecca Redfern, Natasha Powers, Donald Walker and Gaynor Western.

Literature Cited

Antoine DM and Hillson SW (2005) Black Death and health in fourteenth century London. Archaeology International (UCL) 8.

Barber B and Bowsher D (2000) The Eastern Cemetery of Roman London: excavations 1983-1990, MoLAS Monograph 4.

Barber B, Chew S, Dyson T and White B (2004) The Cistercian Abbey of St Mary Stratford Langthorne, Essex: archaeological excavations for the London Underground Limited Jubilee Line Extension MoLAS Monograph 18.

Bentley D and Pritchard F (1982) The Roman Cemetery at St Bartholomew's Hospital. Transactions of the London and Middlesex Archaeological Society 33: 134-172.

Bowsher D, Dyson T, Holder N and Howell I (2007) The London Guildhall: an archaeological history of the neighbourhood from early medieval to modern times. MoLAS Monograph 36.

Brickley M and Miles A (1999) The Cross Bones Burial Ground, Redcross Way, Southwark, London. MoLAS Monograph 3.

Cowan C (2003) Urban Development in North-western Roman Southwark: excavations 1974-90. MoLAS Monograph 16.

Cowie R, Bekvalac J and Kausmally T (2007) Late 17th-19th-century burial and earlier occupation at All Saints, Chelsea Old Church, Royal Borough of Kensington and Chelsea MoLAS Archaeology Studies Series.

Drummond-Murray J, Thompson P and Cowan C (2002) Settlement in Roman Southwark: archaeological excavations (1991-8) for the London Underground Limited Jubilee Line Extension project. MoLAS Monograph 12.

Gilbert M, Thomas P, Cuccui J, White WJ, Lynnerup N, Titball RW, Cooper A, and Prentice MB (2004) Absence of *Yersinia pestis*-specific DNA in human teeth from five European excavations of putative plague victims. Microbiology 150: 341-354.

Gilbert M, Thomas P, Cuccui, J, White, W J, Lynnerup, N, Titball, R W, Cooper Alan and Prentice, M B (2004) Response to Drancourt and Rauolt. Microbiology 150: 264-5.

Gilchrist R and Sloane B (2005) Requiem: the medieval monastic cemetery in Britain. MoLAS (Surveys and Handbooks series).

Gowland RL and Chamberlain AT (2005) Detecting plague: palaeodemographic characterisation of a catastrophic death assemblage. Antiquity 79: 146–157.

Hall J (1996) The cemeteries of Roman London: a Review. In J Bird, M Hassall and H Sheldon (eds.): Interpreting Roman London: papers in memory of Hugh Chapman, Oxbow Monograph 58, pp57-84.

Hiller J and Wilkinson DRP (2005) Archaeology of the Jubilee Line extension: prehistoric and Roman activity at Stratford Market Depot, West Ham, London, 1991-3. MoLAS (Surveys and Handbooks series).

Malcolm G, Bowsher D and Cowie R (2003) Middle Saxon London: excavation at the Royal Opera House 1989-99, MoLAS Monograph 15.

Mackinder A (2002) A Romano-British Cemetery on Watling Street: excavations at 165 Great Dover Street, Southwark, London MoLAS Archaeology Studies Series 4.

Margerison BJ and Knüsel CJ (2002), Paleodemographic comparison of a catastrophic and an attritional death assemblage. American Journal of Physical Anthropology 119: 134-143.

Miles A, White W and Tankard D (2007) Burial at the site of the parish church of St Benet Sherehog before and after the Great Fire: excavations at 1 Poultry, City of London MoLAS Monograph.

Miller P and Saxby D (2007) The Augustinian priory of St Mary Merton, Surrey: excavations 1976-90. MoLAS Monograph 34.

Milne G (1997) St Bride's Church London: archaeological research 1952-60 and 1992-5. English Heritage: London.

Molleson T and Cox M (2003) The Spitalfields Project vol 2: The Anthropology: the middling sort CBA Research Report 86. Council for British Archaeology: York.

Roberts C and Connell, B (2004) Palaeopathology In MB Brickley and J McKinley (eds.): Standards for Human Bone Recording BABAO/IFA, pp. 34-49

Rock WP, Sabieha AM and Evans RI W (2006) A cephalometric comparison of skulls from the fourteenth, sixteenth and twentieth centuries. British Dental Journal, 200: 33-37.

Schofield J and Lea R (2005) Holy Trinity Priory, Aldgate, City of London: an archaeological reconstruction and history. MoLAS Monograph 24.

Sidell J, Cotton J, Rayner L and Wheeler L (2002) The Prehistory and topography of Southwark and Lambeth MoLAS Monograph 14.

Sloane B and Malcolm G (2004) Excavations at the Priory of the Order of the Hospital of St John of Jerusalem, Clerkenwell, London MoLAS Monograph 20.

Swift D (2003) Roman Burials, Medieval Tenements and suburban growth: 201 Bishopsgate, City of London, MoLAS Archaeology Study series 10.

Thomas C (2002) The Archaeology of Medieval London Sutton: Stroud.

Thomas C (2004) Life and Death in London's East End: 2000 years at Spitalfields. MoLAS.

Thomas C, Sloane B and Phillpotts C (1997) Excavations at the Priory and Hospital of St Mary Spital, London MoLAS Monograph 1.

Waldron TA (1992) Osteoarthritis in a Black Death cemetery in London. International Journal of Osteoarchaeology 2: 235–240.

Waldron HA (2001) Are plague pits of particular use to palaeoepidemiologists? International Journal of Epidemiology: 30 104–8.

Watson S (2003) An excavation in the Western Cemetery of Roman London: Atlantic House, City of London, MoLAS Archaeology Study Series 7.

Watson S and Heard K (2006) Development on Roman London's western hill: excavations at Paternoster Square, City of London. MoLAS Monograph 32.

White WJ (1988) Skeletal Remains from the Cemetery of St Nicholas Shambles, City of London London & Middlesex Archaeological Society, Special Paper No 9.

Wood J and Dewitte S (2003) Was the Black Death yersinial plague? Lancet 3: 327-8.

Cracking the Code: biography and (reconstructed) stratigraphy at St Pancras burial ground

Phillip A. Emery

Principal Archaeologist,
Gifford
Pentagon House
52-54 Southwark Street
London SE1 1UN
UK
e-mail address for correspondence: phil.emery@gifford.uk.com

Abstract

Archaeological investigation of St Pancras Burial Ground took place in 2002–3, in advance of construction of the new London terminus for the Channel Tunnel Rail Link. This resulted in the mapping of over 1300 burials within an extension of the cemetery in use between 1793 and 1854. Of these, some 780 individuals were recovered for osteological analysis. In addition, over 1100 items of associated metal coffin furniture were recorded. Documentary evidence indicates that the buried population was remarkably heterogeneous, in religious and social terms, including increasing numbers of paupers as well as prominent citizens and aristocrats (including many Roman Catholics and French émigrés).

This tightly-dated body of evidence has many strands (osteological, biographical, social and art-historical) and offers exceptional potential for finely detailed studies of Londoners' lives and circumstances in a turbulent era. The fieldwork environment was demanding, with archaeologists working under pressure alongside a cemetery clearance contractor. The key challenge was that of maximising the size of the sample of recorded burials while ensuring the integrity of the record, and thus the analytical value of the resulting datasets.

This paper will outline the archaeological methodology which was applied, explaining how it developed during the course of this complex and often controversial project. Although the circumstances defied any rigorously stratigraphic approach, three-dimensional recording of burial locations permitted much detailed stratigraphic reconstruction of the likely inter-relationships between groups of burials, and thus correlation of burials with the detailed biographical and other information provided by the cemetery burial registers. This 'decoding' operation is crucial to the integration of the osteological and documentary information available.

Keywords: St Pancras, 18th/19th century, cemetery, stratigraphy, coffin plates, burial registers

Introduction

In 2002-3 a team of archaeologists attended the partial clearance of St Pancras burial ground by an exhumation contractor under the provision of Schedule 11 of the Channel Tunnel Rail Link Act 1996. Research into the wealth of documentary evidence showed that St Pancras harboured an unusually diverse and tightly-dated buried population, providing rare and important opportunities for demographic and social enquiry into an urban community that was experiencing dramatic growth and change. In addition, once operations had started on the site, preservation of buried remains was observed to be exceptional. It became apparent, especially given the numbers of burials that were to be exhumed, that the conduct and outcome of the St Pancras project were likely to have implications for the approach to future investigation of other post-medieval cemeteries.

The deployment and rôle of clearance firms on post-medieval burial grounds, which have been recognised as being archaeologically significant, persists as a subject of debate (Bashford & Pollard 1998: 165; Roberts & Cox 2003: 289; Boyle 2004: 72). While the framework for treatment of post-medieval cemeteries evolves, it is important that archaeologists continue to engage pragmatically with clearance operations. There are three principal reasons; 1) archaeologists' cumulative experience of adapting to project-specific constraints will crucially shape professional best practice, 2) the interaction of archaeologists and exhumation contractors will help to establish mutual understanding and agreed protocols for a future in which co-existence of the two types of practitioners seems inevitable, and 3) by their active participation, archaeologists signal the value that they place on post-medieval burial grounds and a commitment to realising their research potential.

Given the circumstances of the watching brief at St Pancras, the challenge was to develop a sustainable recording regime that would deliver a baseline of consistency and integrity to allow meaningful analyses of material collected under a wide range of conditions. At the conclusion of analysis the constraints on interpretation resulting from the work not being undertaken as a controlled archaeological excavation can be evaluated (White forthcoming).

Historical Background

St Pancras Old Church occupies a particularly interesting place in London's history. With possible pre-Conquest origins, it has been claimed as the oldest surviving church in London (Figure 1). However, it was only in the 18th century that substantial enlargement of the churchyard, reflecting dramatic population growth, took place. At this time, St Pancras churchyard became a customary place of burial for prominent citizens and aristocrats. Pauper burials were also increasing, however, and it is clear that the people interred formed a very diverse group. In 1793 the churchyard was extended into a new burying ground and the older part of the churchyard closed. It was the southern part of this 1793 extension that was the subject of investigation in advance of works associated with the Channel Tunnel Rail Link in 2002–3 (Figure 2).

Figure 1. Engraving of St Pancras Old Church viewed from the south by Alexander Hogg, 1784.

Long associated with London's Roman Catholic community, St Pancras became the natural resting place for refugees from the French Revolution. As well as aristocrats and their households, the new arrivals included economic migrants, including artisans producing luxury items whose bourgeois market in Paris had collapsed. Further, some 5000 clerics refusing to sign the Oath of Allegiance to the Civil Constitution of the Clergy and fearing deportation to French Guiana also sought asylum. Many settled in Somers Town and Bloomsbury, and 'many hundred' are known to have been buried at St Pancras (Bellenger 1986: 137). The parish of St Pancras lay on the fringe of the growing Metropolis and offered cheap housing (Figure 3). It absorbed successive waves of migrant workers associated with construction of the Regent's Canal in the 1820s and the railways from the 1830s onwards. The burial ground saw very heavy use, and was closed in 1854 with the passing of Acts of Parliament for the provision of municipal cemeteries on the outskirts of London.

Figure 2. View from inside the southern part of the 1793 burial ground extension. The south wall of the cemetery can be seen in the centre of the picture. The burial ground was affected by (left) the deck extension required by Eurostar trains, which are 394m long, and (right) the burrowing junction connecting with the new underground Thameslink station

Disturbance of the burial ground soon followed with the construction of the Midland Railway in the 1860s. This led to public outcry. As a young architecture student working under Arthur Blomfield (clerk of works appointed by the Bishop of London), Thomas Hardy oversaw the exhumation works at St Pancras (Tomalin 2006: 80-81). A burial pit forty feet deep was excavated on the site of the present-day Coroner's Court for the reinterment of the remains of over 7000 individuals. In 1882, some 16 years later, Hardy, by now a celebrated novelist and poet, was moved by his experiences to write a poem entitled 'The Levelled Churchyard', the second verse of which reads:

> We late lamented, resting here
> Are mixed to human jam
> And each to each exclaims in fear
> I know not which I am!

Over 130 years later, a second partial clearance was required to make way for the new London terminus for the Channel Tunnel Rail Link. While it was clear that the station deck extension for the long platforms needed by Eurostar trains would have a major impact on the southern

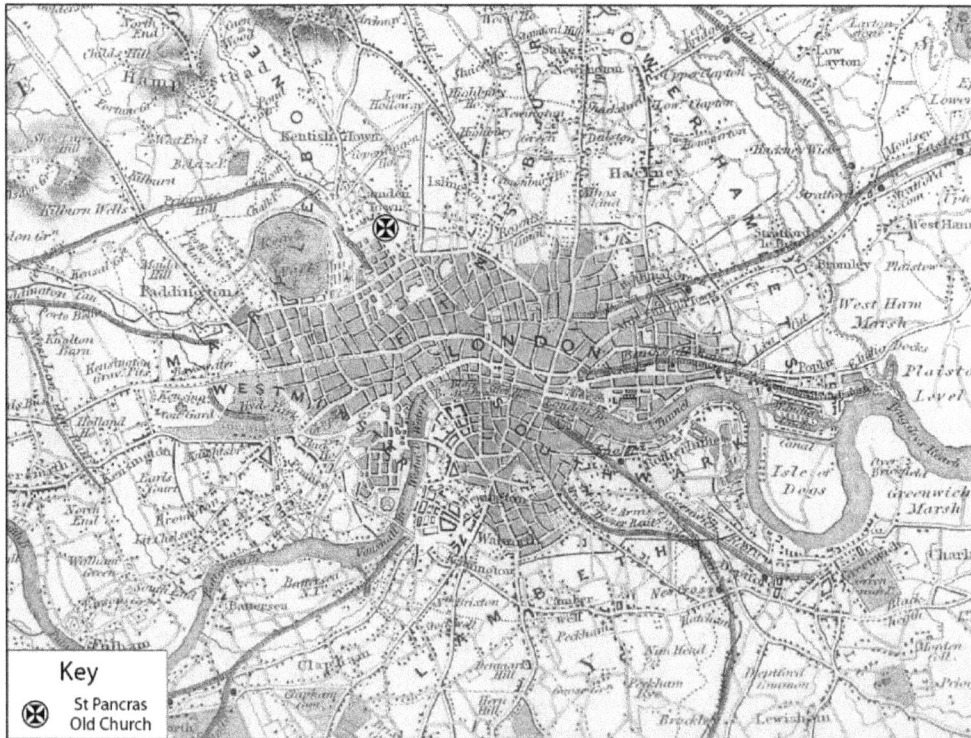

Figure 3. Thomas Moule's map of London, 1836 (detail) showing the location of St Pancras church. This also provides a snapshot of early railway developments on the outskirts of the city, including the London & Birmingham Railway terminating at Euston.

part of the burial ground, this area could not be evaluated archaeologically in advance of groundworks due to the presence of the operational railway both above and below ground. From February 2002 to June 2003, control of excavations within the burial ground was assigned by the project managers, Rail Link Engineering, to a cemetery clearance contractor, Burial Ground Services UK Ltd (BGS). Gifford was appointed as the archaeological organisation to undertake an archaeological watching brief on the cemetery clearance works, working with a field team from Pre-Construct Archaeology (PCA) led by Kevin Wooldridge, specialists from the Museum of London Specialist Services and documentary researcher Chris Phillpotts.

Rules of engagement

Archaeologists had to work within the demands of the project timetable, working in two shifts to cover 14-hour days and using artificial lighting when necessary. Access restrictions at various stages during the works often governed the level of archaeological recording. At first, the exhumation contractors' excavations were a series of machined sondages up to 3m wide and 3m deep, separated by 2m-wide baulks, which were cleared individually to full depth (Figure 4). This methodology, designed to safeguard the nearby Midland Main Line railway and Thameslink Tunnel which remained in use, severely limited the safe working space available for exhumation operatives and archaeologists, and hence inhibited archaeological retrieval.

Figure 4. Two stacks of coffins from adjacent rows briefly exposed in the side of a machine-excavated sondage

These difficulties were understood by all parties, and therefore two small areas in the southern end of the cemetery were set aside for detailed archaeological recording. Controlled excavation of 83 inhumations (all of them subsequently studied osteologically) in this sample area permitted direct stratigraphic recording by archaeologists, and careful definition of some of the relatively flimsy and poorly preserved coffins in the upper levels of the cemetery. During the later years of the cemetery's use, when overcrowding had reached extreme proportions, coffins had been laid side by side, head-to-toe, in long burial trenches, apparently to make best use of the available space.

Figure 5. Exhumation in progress, with soil being removed mechanically in spits

This approach to archaeological recording was a pragmatic response to the demanding conditions under which exhumation was taking place: taking control of works in specific areas to ensure that a meaningful archaeological record was made had come at the cost of partial disengagement from the wider exhumation works. Yet even this compromise approach was to prove unsustainable. Under pressure from the hectic project schedule during Autumn 2002, archaeological site recording was suspended on 15 November and the sample excavation abandoned. For two days exhumation continued mechanically, unattended by archaeologists. However, this development aroused concern in many quarters, including the Church of England, English Heritage, the Council for British Archaeology, the British Association for Biological Anthropology and Osteoarchaeology, and Rescue, as well as the general public and news media. It became clear to all parties, including the client and contractors, that archaeologists had to be involved with the exhumation process on some meaningful basis, and discussions followed.

The project management proposed certain rules of archaeological engagement. All excavation and lifting of human remains was to be undertaken by cemetery clearance operatives; archaeologists would not clean coffins or their contents, and were to make their record

from the excavation edges. It was clear, however, that these conditions would make it difficult to deliver any coherent and meaningful record or to guarantee the provenance and integrity of osteological samples. Gifford argued that the cornerstone of any summary recording procedure was accurate three-dimensional surveying of burials. This might provide some basis for reconstructing the broad stratigraphic sequence of events in the cemetery, offsetting the lack of opportunities for hands-on recording, but could only be achieved if archaeologists were allowed into the excavation area itself.

Acknowledging these issues, Rail Link Engineering and the construction contractor agreed that an archaeologist would be allowed to survey four corner points on a coffin if its occupant was being recovered for osteological study, and two points if it was not. The on-site procedures, showing how archaeological recording fitted around the exposure and opening of coffins and the bagging of their contents by BGS, were articulated in a simple flow chart. Substantial adjustments to the construction programme were made to accommodate the outstanding works, and exhumation operations finally resumed on 6 January 2003. The archaeological watching brief and exhumation works were both now made easier by a change in exhumation method, the contractor being allowed to machine soil in a series of shallow spits as originally envisaged (Figure 5).

Results

Approximately 1300 burials were recorded three-dimensionally with a total station theodolite under the supervision of Duncan Sayer (PCA) (Figure 6). Establishing the relative positions of individual burials has proved to be of huge significance, both in reconstructing the stratigraphic sequence of coffins and correlating inhumations with entries in the burial registers. For the period 1793–1804, most entries in the parish burial registers offer an alphanumeric grave plot reference. During analysis, the fieldwork director Kevin Wooldridge 'decoded' this system by matching the positions of a sufficient number of named burials with register entries to allow the identities of almost a hundred 'anonymous' burials to be inferred.

Certain inherent site characteristics affected the quality of the evidence and integrity of the recorded sample. For example, the clay soil enhanced preservation of coffins but often impeded archaeological work during the winter weather. Often multiple coffins were stacked within a single grave (Figure 4), with the burial registers indicating successive (often unrelated) interments in a single plot within only a few days. Where coffins had broken, there was the potential for bones to have been commingled and mixed between individuals. In many instances, ground pressure had caused the sides of the coffins to slump and twist inward, thereby compounding these problems.

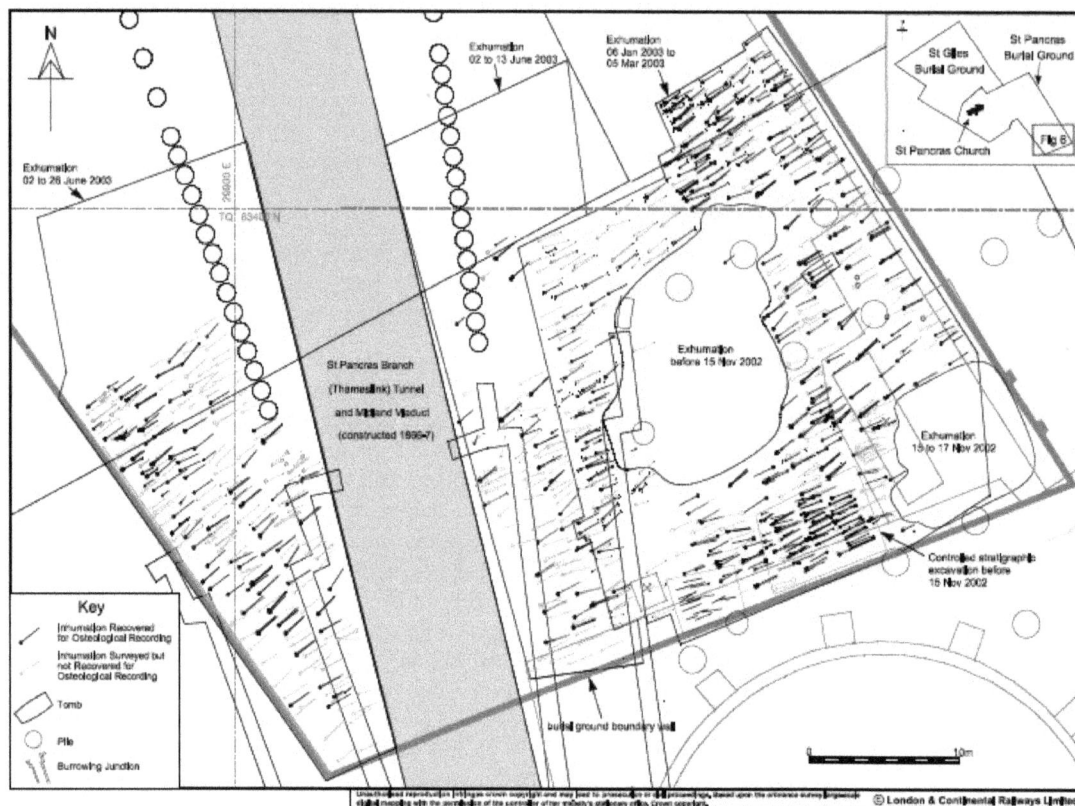

Figure 6. Plan of burials and funerary remains recorded in 2002-3 in the southern part of the 'Third Ground' (part of the 1793 extension to St Pancras churchyard), showing the principal elements of the CTRL terminus development that affected the cemetery.

Bone from skeletons recovered for osteological study was in excellent condition. Some of the bodies still possessed hair on the head, and even eyebrows (White forthcoming). Grave cuts excavated into the clay acted as sumps, so coffins at the base of each were usually found to have been permanently filled with water. The resulting anaerobic preservation led to remarkable survivals of floral tributes (Figure 7) and grave clothes, but could be a mixed blessing when it resulted in preservation of human soft tissue. More than 30 skeletons still had soft tissue present, chiefly brain matter but one had flesh adhering to the arms (White forthcoming). For reasons of health and safety, it had been agreed that this would be subject to palliative treatment by the exhumation contractor, precluding retention of some of the remains for osteological study.

A wide variety of pathological conditions were visible in the sample of skeletons selected for analysis, including those considered as endemic to urban populations of this period, such as syphilis (Powers & Emery, this volume). Also of particular interest at St Pancras is the comparatively large group of sub-adults (183 individuals) recovered for analysis, which has allowed comparison of child growth and development with other sites.

Considering the full analytical range of this investigation, the single most valuable facet of the evidence was the exceptional preservation of the decorative metal coffin fittings. The collection of c.1100 fittings was significant not only in art-historical terms, but because it included over 150 inscribed breast-plates indicating the name (and, by implication, sex), age and date of death of the deceased (Figure 8). Their manufacture from thin iron sheet, sometimes dipped in tin, makes their legibility still more remarkable. At the post-excavation stage, it was the association of these legible inscriptions with individual inhumations that allowed integrated study of human osteology and documentary evidence on an unprecedented scale for a flat (i.e. non-crypt) cemetery.

Unsurprisingly, many of the inscriptions were in French. Two senior emigrant clerics were identified; Pierre Augustin Godart de Belbeouf, last Bishop of Avranches, and Arthur Richard Dillon, Archbishop of Narbonne and Primate of Languedoc (Figure 9). Dillon had been provided for in retirement by his cousin, Viscount Dillon, who lived in London, and had been buried with his fine set of porcelain dentures (Figure 10). Natasha Powers (Museum of London) has explored their origins, concluding that these are likely to be the work of Nicholas Dubois de Chemant, who had a furnace at the Sèvres factory and obtained a royal patent for 'mineral paste teeth' from Louis XVI in 1789, before leaving for England as an economic migrant in 1792 (Powers 2006).

Figure 7. Remains of a floral tribute comprising leaves and twigs of box and bay.

Figure 8. A typical engraved coffin breastplate

As the sampling of burials for osteological study continued on site, skeletons which had been analysed, re-bagged and labelled were returned in batches to the exhumation contractor for immediate reburial at St Pancras and Islington Cemetery, East Finchley.

Conclusions

The treatment of post-medieval burials encountered during developments continues to arouse debate within the archaeological profession and beyond. The issues under discussion involve highly sensitive matters of care and propriety, as well as consideration of the academic value of data recovered from them by archaeologists. The conduct and the outcome of the St Pancras project have highlighted a wide range of these issues. Lessons learnt at St Pancras, a flat cemetery, complement those so candidly acknowledged following the investigation of the crypt of Christ Church, Spitalfields (Reeve and Adams 1993, 130).

Figure 9. Portrait of Arthur Richard Dillon, Archbishop of Narbonne (b.1721, d.1806), painted by John Hoppner in 1800.

11

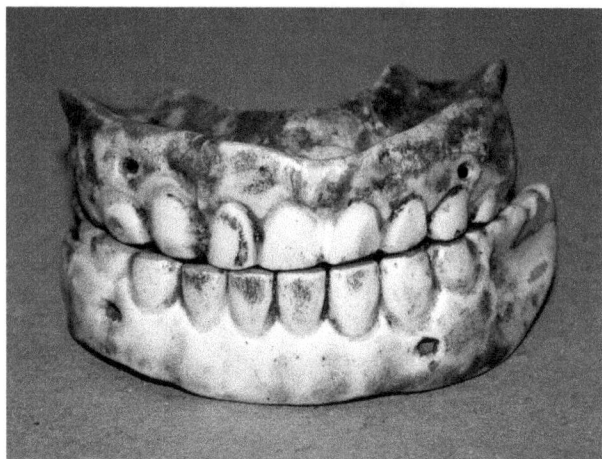

Figure 10. The porcelain dentures recovered from the coffin of Archbishop Dillon

The use period of the 1793 burial ground extension saw London transformed, with massive population growth and immigration accompanied by an industrial revolution and by far-reaching changes in people's lives, including a rise in urban poverty. The recorded St Pancras burials, forming a large, tightly-dated collection of bone, in generally excellent condition, represent an extremely important sample of London's population in this turbulent period. The St Pancras assemblage provides a 'snapshot' of the health of Londoners in the late eighteenth and early nineteenth centuries. This is a notable period of population increase, industrialisation and a subsequent increase in the volume of urban poor (Powers and White forthcoming). Analysis of the bone, which amounts to a significant population-based osteological study for a post-medieval non-crypt assemblage, has now been completed by Bill White and Natasha Powers at the Museum of London, assisted by Don Walker, James Langthorne and Kathelen Sayer of PCA. The degree of correlation between individual skeletons and the burial registers has permitted the creation of fascinating 'biographies' for some of the deceased, and the results may be set against demographic, census and mortality data from other sources. With reference to her seminal investigation of the burials in the crypt of Christ Church, Spitalfields, Reeve has reminded us that the examination of human remains from this period is essential to complement documentary information about disease, the history of dentistry and medicine, demography and the local population, and also to genealogical studies (Reeve 1997: 5).

However, the significance of the St Pancras project is also methodological and political. Developments involving major exhumations will continue to challenge archaeologists in the future; St Pancras offers a powerful case study illustrating the hard decisions that must sometimes be made within the context of a large and fast-moving project. It emphasises the importance of not only collecting stratigraphic and osteological information and sample material, but also in understanding early in the process how that information might need to be used during analysis. Here, archaeologists' clarity during their

dialogue with developers as to which aspects of recording were of greatest important to them was crucial, not only to reconstruct the stratigraphic development of the burial ground during analysis, but also to assign identities to many recorded burials of individuals who would otherwise have remained nameless.

Acknowledgments

The archaeological investigation at St Pancras burial ground formed part of an extensive programme of works along the route of the Channel Tunnel Rail Link funded by London & Continental Railways. The author wishes to extend particular thanks to Helen Glass (Archaeology Manager for Rail Link Engineering), Kevin Wooldridge (Fieldwork Director) and his team of dedicated archaeologists from PCA, specialists from the Museum of London and documentary researcher Dr Chris Phillpotts.

Literature Cited

Bashford L and Pollard T (1998) In the burying place – The excavation of a Quaker burial ground. In M Cox (ed.): Grave Concerns: Death and Burial in England 1700-1850; CBA Research Report 113, York: pp154-166.

Bellenger DA (1986) The French Exiled Clergy in the British Isles after 1789. Bath: Downside Abbey.

Boyle A (2004) What price compromise? Archaeological investigations at St Bartholomew's Church, Penn, Wolverhampton. Church Archaeology Volume 5 & 6 p69-79.

Emery PA and Wooldridge K (forthcoming) St Pancras Burial Ground: archaeological investigations undertaken as part of the development of the Channel Tunnel Rail Link's new London terminus. Gifford Monograph.

Powers NI (forthcoming) In Emery, P.A and Wooldridge, K. (eds.): St Pancras Burial Ground: archaeological investigations undertaken as part of the development of the Channel Tunnel Rail Link's new London terminus. Gifford Monograph.

Powers NI (2006) An eighteenth century porcelain dental prosthesis belonging to Archbishop Arthur Richard Dillon. British Dental Journal 201: 459-463.

Powers NI and Emery PA 'The 'French Disease': Syphilis and the burial ground of St Pancras'. This volume.

Reeve J and Adams M (1993) The Spitalfields Project, volume 1 The Archaeology: across the Styx, CBA Research Report 85, York.

Reeve J (1997) Grave expectations. Building Conservation Directory Special Report (The Conservation and Repair of Ecclesiastical buildings), 4-6.

Roberts CA and Cox M (2003) Health and disease in Britain: Prehistory to the present day. Stroud: Sutton Publishing Ltd.

Tomalin C (2006) Thomas Hardy: the Time-Torn Man. London: Viking.

White W (forthcoming) In PA Emery and K Wooldridge. (eds.): St Pancras Burial Ground: archaeological investigations undertaken as part of the development of the Channel Tunnel Rail Link's new London terminus. Gifford Monograph.

White W and Powers NI (forthcoming) In PA Emery and K Wooldridge. (eds.): St Pancras Burial Ground: archaeological investigations undertaken as part of the development of the Channel Tunnel Rail Link's new London terminus. Gifford Monograph.

Recent Excavations in the 'Southern Cemetery' of Roman Southwark

Melissa Melikian[1*] and Kathelen Sayer[2]

[1] AOC Archaeology Group
Unit 7
St Margarets Business Centre
Moor Mead Rd
Twickenham
TW1 1JS

[2] Pre-Construct Archaeology Ltd
Unit 54 Brockley Cross Business Centre
96 Endwell Road
Brockley
London
SE4 2PD
*e-mail address for correspondence: melissamelikian@aocarchaeology.co.uk

Abstract

At America Street in Southwark, London, antiquarians first recorded the presence of Roman graves during the digging of the foundations for a public house in the 19th century. Excavations in 2001-2 by AOC Archaeology Group confirmed these observations by unearthing a total of 151 burials preserved beneath the cellar of a warehouse, the largest Roman cemetery cluster to be yet found in Southwark. The majority of the burials were of adults interred in coffins between the mid 2nd and late 3rd centuries. Subsequently in 2004, an excavation at Lant Street, located 400m to the south of America Street, revealed eighty-eight inhumations and two cremation burials, as well as ditches and linear features defining discrete areas of burial as well as a ritual well or shaft.

Numerous grave goods were found with the burials on both sites, including complete pottery and glass vessels, hob-nailed boots and a significant amount of personal jewellery. Notable finds include finger-rings with engraved intaglios, hundreds of jet beads, a casket with copper fittings and decorative bone inlay and a folding knife with an ivory handle carved in the form of a leopard. A number of interesting burial practices were noted. At America Street, for example, these included a horse skull placed in one burial, decapitation, possible family groups and weaponry, whilst at Lant Street these included a high number of prone burials, a linear feature with multiple burials and a dog burial with associated grave goods.

Keywords: Roman Cemetery, London, Burial Practice, Demography

The Archaeology of Roman Southwark

The City of Southwark is situated within metropolitan London, immediately south of the Thames and the City of London. The nature of the Roman settlement in Southwark was dictated by the local topography. At that time, Southwark was predominantly marshy with two low islands of clay-capped sand and gravel. The area of the Roman settlement concentrated on these northern and southern islands, divided by what is known today as the 'Southwark Street Channel'.

At this point Southwark, on the southern side of the Thames, provided the first suitable crossing point, and later a bridge, which was accessible by road from the major invasion entry points on the coast. The northern island of Southwark was the only dry land that did not flood at High Tide and as such provided the shortest crossing point across the Thames. The southern bridgehead itself has not been discovered but is inferred from the two major roads which converged at the riverbank (Sheldon 2000: 129-131; Yule 1988: 15-16). The nature of the settlement in Roman Southwark, however, is difficult to establish. During the early Roman period, before major land reclamation had occurred, Southwark would not have been a desirable area for habitation; it was situated on marginal land external to the *civitas* (Yule 1988). Occupation was initially trade and industry based, situated alongside one of the major roads into *Londinium*. Evidence for the processing and production of foodstuffs, bone objects, metalworking and leatherworking have all been found in the area (Sheldon 2000: 140-142; Drummond-Murray & Thompson 2002). It appears that during the Roman period the character of Southwark changed from that of an early trading and industrial centre to later incorporating an administrative and military presence (Yule 2005). The excavation of such recent sites as the 'temple complex' discovered at

Tabard Square (Killock, forthcoming) and the excavation of the two major cemeteries, at America Street and at Lant Street, are now increasing the knowledge and interpretation of Roman Southwark.

Roman Burial Customs

Roman law decreed that a body was not to be buried or burned within the city. The reasoning behind this was twofold; a sanitary precaution and the fear of defilement (Toynbee 1996: 43). The Romans, fearful of evil spirits, took great care in the burial of their dead. The dead were not forgotten but were visible to the living population and cemeteries developed alongside the major roads beyond the city gates. For the first two hundred years, the predominant method of burial in Roman Britain was cremation. In the mid-second century, the practice of inhumation was introduced from the continent. The reason for the change is uncertain, whether it was because of changes in religious beliefs or merely fashion. Inhumation became increasingly popular with time and, by the fourth century, was the generally accepted method of burial. Roman burials commonly contained grave-goods (following the pre-Christian tradition). Traditionally, these included jewellery and personal ornaments, eating and drinking vessels and coins. The purpose of these grave-goods was to honour the deceased as well as to equip them in death, both in the grave itself and for their journey to the underworld (Alcock 1980, 56-62). Romano-British burial practice was not solely Roman but a synthesis of Roman and Celtic beliefs and traditions; certain burial practices have their roots in the Iron Age and others were a newly adopted Mediterranean tradition.

The Cemeteries of Roman London

By *c.* AD 100 large urban cemeteries had been established around the perimeter of Roman London. These burial areas clustered in four distinct groups; to the north, east and west of the northern settlement and to the south of the Thames in Southwark, although this burial area does not conform to those on the north bank (Barber & Hall 2000: 109).

The Northern Cemetery

The northern cemetery flanked the Roman road Ermine Street, which connected London to Lincoln. The road led out beyond Bishopsgate and burials have been recorded from the 16th century onwards in addition to a number of more recent excavations in the area. To date, these burials total 28 cremation burials and 181 inhumations (Barber & Hall 2000: 108-109). The cremation burials span the 1st to 4th century while the inhumations date from AD100 onwards but mainly date to the 3rd and 4th centuries. The excavations at Spitalfields fall within this later group (Thomas 2004).

The Eastern Cemetery

The eastern cemetery is by far the largest Roman cemetery to be excavated from London. Eleven separate excavations took place in the eastern cemetery in the late 20th century and the cemetery is believed to have covered an area of approximately 12ha (Barber & Bowsher 2000: 1). Located immediately to the east of the city wall outside Aldgate, the cemetery was in use from the 1st to the 5th centuries. Here burials totalled 157 cremation burials and 550 inhumations (Barber & Hall 2000: 110-112).

The Western Cemetery

The western cemetery covers an area from Smithfield to Ludgate. Evidence for the cemetery has been found between the Roman gates of Newgate and Aldersgate close to the city wall and a few burials in the area of Ludgate. Major cemetery excavations in this area include those at Giltspur Street (MacLaughlin & Scheuer unpublished summary report), St Bartholomew's Hospital (Bentley & Pritchard 1982) and Atlantic House (Watson 2003). A total of 59 cremation burials and 189 inhumations have been discovered in the area to date (Barber & Hall 2000: 107-8).

The Southern Cemetery

Unlike the cemeteries to the north of the settlement, the extent of the southern cemetery is difficult to define due to the expansion and contraction of the Roman settlement. It is unclear if the burials form one cemetery or are disparate groups as they do not seem to conform to the layout of the other extra-mural cemeteries. Prior to the excavations at America Street and Lant Street, Roman burials have been predominantly found in two areas of Southwark (Hall 1996: 74-83; Barber & Hall 2000: 104-7). The first is in the area south of the junction of Stane Street and Watling Street near St George's Church, Borough. The cemetery at Great Dover Street falls into this group (MacKinder 2000). A second group lies further north towards the river, sited along the road that is thought to run from the Southwark bridgehead to Lambeth. Prior to the excavations at America Street and Lant Street, only 43 cremation burials and 73 inhumations had been recorded from Southwark (Barber & Hall 2000: 104-107).

Following an evaluation in October 2001, an open area excavation was conducted by AOC Archaeology Group on land adjacent to 1 America Street (NGR TQ 3220 8010) between December 2001 and March 2002. The site was an approximately rectangular plot of land measuring *c.* 21m (east-west) by *c.*12m (north-south), an area of 288 square m. What follows are the preliminary results of the post-excavation work. The full results will be published as a future monograph (Melikian forthcoming).

Examples of 151 burials were excavated at the site, including 16 multiple burials, totalling 167 individuals, 163 of which were inhumations and 4 were cremation burials (Fig 1). The cemetery at America Street dated from the 2nd to 4th centuries. Like areas of the eastern cemetery, the level of inter-cutting at America Street suggests that land was at a premium and, for this reason, over three quarters (77%) of the burials were truncated.

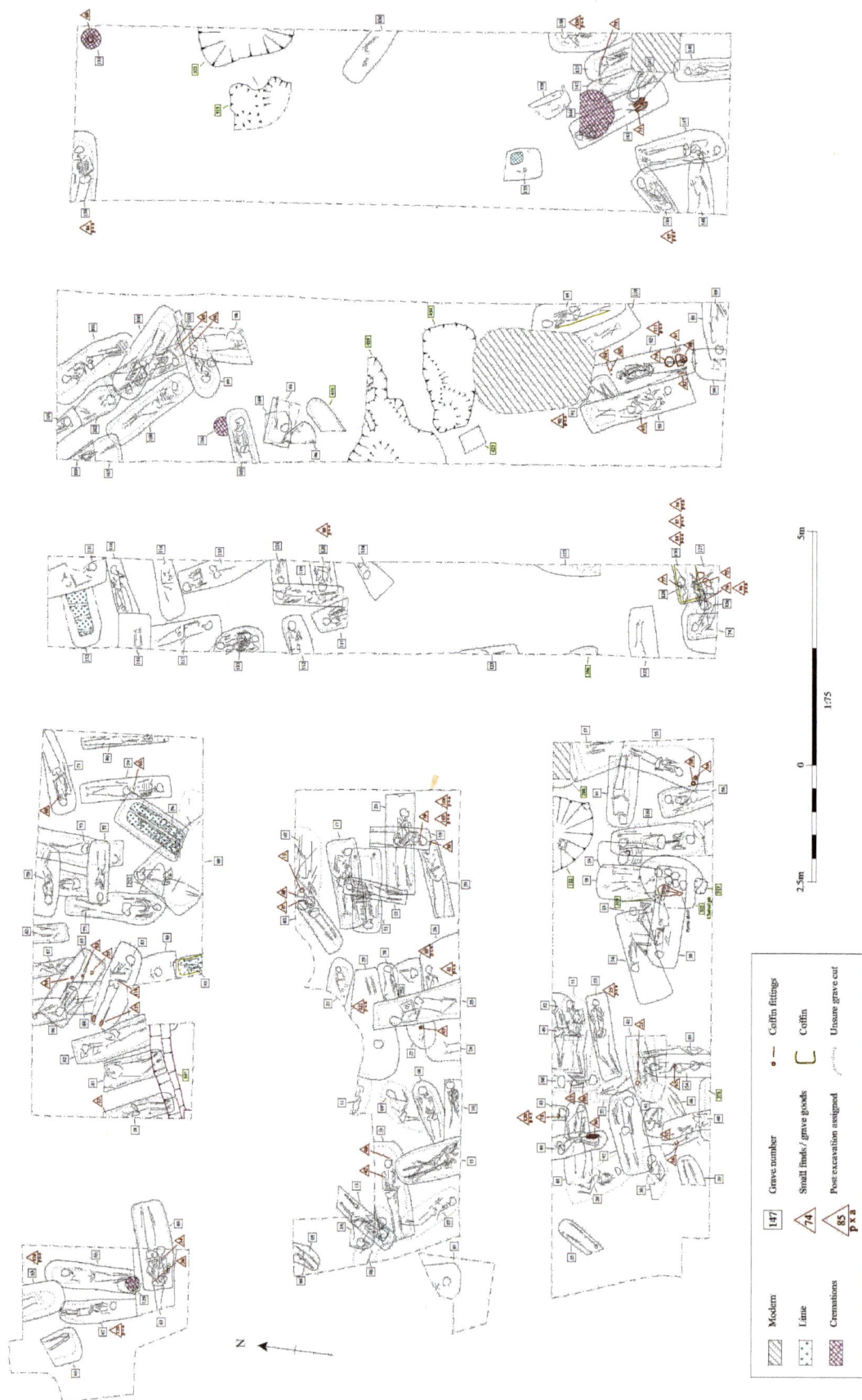

Figure 1. America Street – The Cemetery

Alignment and orientation

The inhumation burials were on a number of different alignments. The most common alignment was north-south (43%) while 38% of burials were on an east-west alignment. This is in contrast to the eastern cemetery where 52.2% were aligned east-west and 44.4% north-south (Barber & Bowsher 2000: 84).

Body position

The majority of inhumations (87%) were supine and extended, which tends to be the most usual practice in urban Roman cemeteries of this date. Approximately 4% of the inhumations were prone; this is comparable with figures from the eastern cemetery (3.3%). More unusual for this type of cemetery is the high number (5%) of crouched burials (Fig. 2). In the eastern cemetery, of the 550 inhumation burials, no crouched burials were recorded (Barber & Bowsher 2000: 87): crouched burials tend to be more common on rural sites.

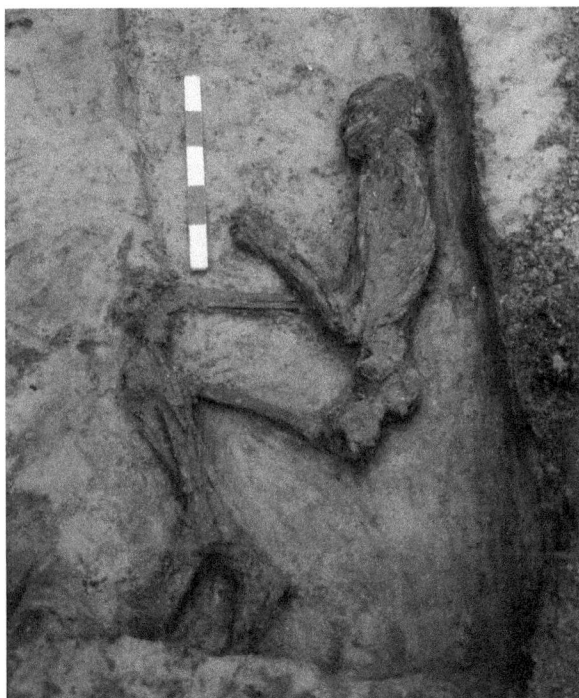

Figure 2. Crouched burial

Coffins

Remains of wooden coffins were preserved in three burials while the fills of 66 graves contained coffin nails in varying quantities. This equates to nearly half (40%) of inhumations being buried within coffins. This is a lower percentage than that at the eastern cemetery where 68.1% of individuals were buried in some type of coffin (Barber & Bowsher 2000: 97).

Chalk burials

A total of five inhumations (3%) had a chalk-like substance packed around the body. This is a lower percentage than that in the eastern cemetery where 12.4% of inhumations were buried in chalk (Barber & Bowsher 2000: 101). At America Street, the substance was found both above and below the body, suggesting when deposited it was originally in liquid or powder form. The chalk-like substance was sampled and analysed to determine whether it was either lime burning waste or decayed natural chalk. The micromorphology and isotopic data strongly suggested a preponderance of natural chalk rather than recarbonated calcium carbonate. It is unclear why chalk was used as a part of Roman burial practice. Philpott (1991: 223) suggests that chalk was a substitute for plaster burials conducted in North Africa and Barber & Bowsher (2000: 321) suggest it may have held a symbolic value. There are numerous examples in all of Roman London's cemeteries. In the eastern cemetery, for example, 81 such burials were identified (Barber & Hall 2002: 112). It could be that there was a belief that the presence of chalk would preserve the body.

Multiple burials

Figure 3. double chalk burial

There were a total of 16 multiple graves. One burial contained the remains of 3 individuals; an adult male, adult female and a juvenile. The male and female appeared to be embracing one another with the juvenile lying in between. DNA analysis was carried out on the individuals in an attempt to ascertain whether a family relationship existed but, unfortunately, the DNA was too degraded to achieve a positive result. One burial consisted of two individuals buried in the same grave-cut but in separate chalk-packed coffins (Fig 3). Interestingly these individuals were of the same approximate biological age; both were 6-7 years old at death. Could it be these were twins who died of the same illness? There were also several stacked burials at the site, containing two or three individuals. The reason for this may be either family groups or to maximise the use of the land.

Ritual activity

A number of burials exhibited unusual and possible ritual burial practices. An iron spear was found in one burial. It is the first weapon to be found in a London cemetery and Philpott (1991) cites no examples of spears or other weaponry in his survey of burial practice in Roman Britain. However, this overlooks the Romano-British double burial from Canterbury which contained two long swords (*spathae*) (Goodburn 1978: 468-471). The practice of burials accompanied by weapons is traditionally an Iron Age phenomenon (Taylor 2001). Apart from its rarity, the unusual feature about the spear is its position. It was found upright, driven vertically between the feet and was found *in situ*. One is tempted to suggest that it could have been a grave marker, but there are many possible explanations for its presence and the ritual involved can only now be surmised.

One burial contained a horse cranium located underneath the shins of an adult male, at 90° to the legs (Figs 4). Iron nails present in the grave fill suggest the individual was buried within a coffin But it is unclear whether the horse cranium was placed within the coffin or outside. The horse was aged 7-11 years at death, making it old enough to have been broken and trained to harness or riding, and thus probably within the best years of its working life. It is interesting that the skull lacked its mandible which may have been removed prior to placement in the burial or perhaps the skull was an object of special significance. The presence of the horse skull is rather enigmatic. Horse remains have been found at other cemetery sites but are generally not found in graves. The only British parallel for the deposition of a horse skull within a burial comes from outside of London, at Alton, where a cremation contained an unburnt horse skull (Philpot 1991: 197). A horse skull is unlikely to represent food and is more likely to be a votive deposit and may be the result of both Celtic and Roman influences (Green 1983).

One burial contained the articulated remains of a juvenile aged 11 years at death with the articulated skull of a six year old placed on the knees (Fig 5). The skull was examined and no cut marks were visible on the bone. Evidence of a coffin, in the form of nails, was found in the grave fill. Cases of decapitation are relatively common in Roman burials; the head may be placed back in the correct anatomical position, on the chest or pelvis, between the legs, or near the feet. There are a number of theories which try to explain this practice; a fear of evil spirits, criminal execution or religious sacrifice (Taylor 2003). However, it is very unusual to have the decapitated head of another individual in a burial.

The grave goods

Grave goods were recovered from 39 inhumation burials

Figure 4. Burial with a horse cranium

Figure 5. Decapitated skull burial

and one cremation, or just under a quarter (24%) of the total number of individuals excavated. This percentage is very similar to the overall figures seen in other cemeteries of *Londinium*. Jewellery or other personal adornment was found in 24 burials, making this the most commonly found category of grave goods. The range of jewellery is limited, but most pieces are typical of late 3rd to 4th century fashions, thus helping to date the burials. Five necklaces were found in the neck or chest areas, suggesting that they were worn. The necklaces were composed of multiple beads either in glass or jet. All individuals found wearing necklaces were female. Bracelets were also worn, again reflecting the fashions of the 3rd and 4th centuries. Evidence for five worn bracelets were found, these consisted of two made of jet, and three of copper alloy. Jet was considered to be beneficial in female burials as it was regarded as having electrostatic properties to ward of any evil spirits (Allason-Jones 1996: 15-17). Finger rings are good examples of personal possessions and three, two of which are 3rd century types, were apparently being worn. The most unusual item of jewellery was a pendant made of silver and blue glass, possibly part of a necklace or a pendant earring.

The custom of placing shoes within a burial is well known from Roman Britain, although its exact significance is subject to debate; footwear was provided for the journey to the underworld or for use in the afterlife. Eight burials contained footwear, in all cases represented by hobnails from nailed shoes or boots, as no leather survived on site. Most of the clusters of nails have also been disturbed and it was generally not possible to recover the nailing patterns. In one burial, the nails were comparatively well preserved and the nailing pattern suggests that the male occupant wore the heavy boots known as *caligae*. Coins were recovered from five adult burials. The placing of a coin in the mouth or on the eyes stemmed from the classical tradition for payment on Charon's barge across the River Styx (Alcock 1980: 57). This practice was adopted by the Romano-British and penetrated both town and countryside and the wealthy as well as the poor. Eight inhumations were furnished with whole or nearly-complete pottery vessels.

Cremations

One urned and three unurned cremation burials were identified from the site. The urned cremation was within an Alice Holt Farnham ware everted-rimmed jar dating to AD 250-400. Two of the cremation deposits contained charred barley, grasses and species from the pea family. The lack of edible foods in comparison to other sites, such as Great Dover Street (Giorgi 2000: 65-66), indicates that they were more likely to have been part of the fuel rather than actual offerings. The human bone from the cremation deposits is currently undergoing analysis.

Demography

A high level of inter-cutting and truncation on site had implications for the level of analysis that could be carried out. Generally the remains displayed poor to moderate levels of surface preservation; 55% were in moderate condition and 36% were poorly preserved. At America Street, three quarters (77%) of individuals were adult and a quarter (23%) were juvenile. Again this is comparable to the eastern cemetery where 70.1% were adult and 23.5% were juvenile (Barber & Bowsher 2000: 311). There were no juveniles aged below six years at death present in the assemblage. This is quite common for Roman cemeteries as they may have been buried in separate areas or simply not regarded as full members of society. In a funerary context the Romans appear to have had a different attitude towards children and Roman law allowed newborns to be buried at their parents' homes (Barber & Bowsher 2000: 313). The number of individuals for whom a sex estimation could be derived was limited by the high level of truncation on site and poor preservation. Half of the assemblage could not be sexed because the relevant diagnostic parts were not present. For the remainder, 30% were male, 16% female and 3% were of intermediate sex (i.e. could neither be classed as male or female). The male to female ratio here was 1.7:1 and is comparable with other urban Roman cemeteries; 1.7:1 at the eastern cemetery (Barber & Bowsher 2000: 311), 1.5:1 at Giltspur Street (MacLaughlin & Scheuer unpublished summary report) and 1.6:1 at Lankhills (Clarke 1979: 123). Examples of more common conditions (degenerative joint disease, periostitis, fractures, a post-rachitic individual) were identified as well as several interesting cases of pathology. One individual, a young adult male, had three linear healed wounds present on the right parietal. A male aged 26-35 years at death exhibited pathology consistent with poliomyelitis; the right leg exhibited atrophy and was substantially thinner and lighter than the left. Previously only five possible instances of poliomyelitis had been described from Roman Britain (Roberts & Cox, 2003).

Lant Street

Excavations carried out in 2004 at Lant Street (NGR TQ 3225 7970) revealed further, similarly dated, inhumations representing either a separate cemetery or a continuity of the burials seen to the north at America Street. A total of 89 inhumations and two cremation burials were excavated in three discrete areas (south, centre and north) of the site, each area separated by ditches or gullies (Fig 6). The vast majority of the inhumations came from the central area. As with America Street, the results presented here represent only the preliminary work: further research and the full results will be published in a forthcoming monograph (Sayer forthcoming).

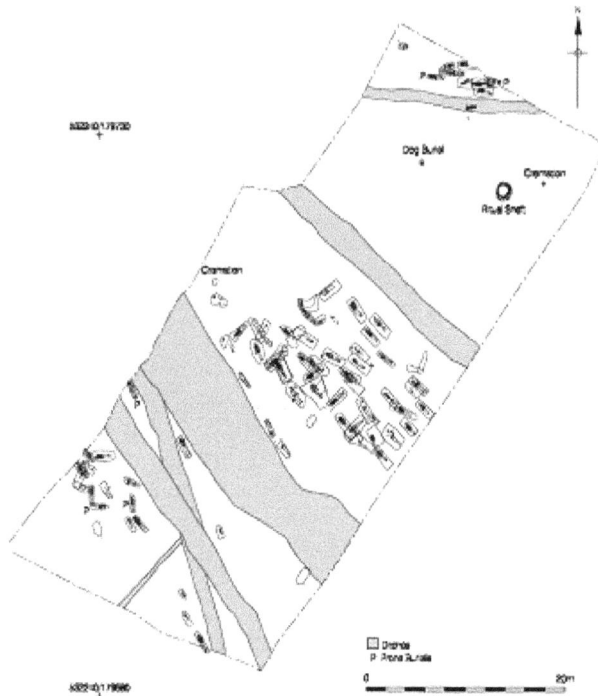

Figure 6. Lant Street – the cemetery.

Alignment and orientation

The inhumations largely respected the alignments of the ditches and gullies, lying either parallel or perpendicular to them. The most common alignment was north-west to south-east (65%), with only equal numbers (15%) orientated east-west, and north-east (14%) to south-west while only a few lay north-south (6%). This contrasts with the majority of the alignments recorded at America Street (see above) and the eastern cemetery (Barber & Bowsher 2000, 84), the former, north-south and the latter, east-west. This was possibly as the result of the alignment with the cemetery with Stane Street, or other features within the landscape, reflecting the local topography.

Body position

The majority of the inhumations (90%) were positioned supine and extended, whilst two skeletons were supine with the legs slightly flexed. A single female was found lying on her side without the careful positioning seen in other burials. In addition, five prone burials were found (6%), contrasting with the number (3.3%) found within the eastern cemetery (Barber & Bowsher 2000, 87). The prone burials, found within the southern and northern burial areas, comprised three young adult males, one young adult female and one middle-aged adult female. It may, therefore, be possible to suggest a tendency for prone burials to be found on the edge or just outside the boundaries of cemeteries and that these areas within the Lant Street site were regarded as being outside the main

burial area. None of the prone burials were buried in coffins, unlike those recorded in the eastern cemetery, 71% of which were coffined (Barber & Bowsher 2000, 87). Further research into the significance of these burials is required.

Coffins

The majority of burials at Lant Street were buried in coffins, as is indicated by coffin nails within 71 of the burials (80%) and also small fragments of mineralised wood from two of these. This is in contrast to the 40% of burials found within coffins at America Street.

Chalk burials

As with America Street and the eastern cemetery, few burials (6.7%) contained the chalk-like substance.

Multiple burials

At Lant Street, only one burial contained more than one individual (Fig. 7). The grave was found in the northern burial area and contained the remains of a female adult and two juveniles, one aged between 4 and 5 years at death and the other an infant of about 9 months. The adult was lying supine and extended with the infant lying over her pelvic region; the older child had been placed at the foot end of the grave. Included within the burial were three pottery vessels; one to the side of the adult's head and one to either side of the infant's head.

Figure 7. Multiple burial

Ritual activity

A small dog wearing a collar was buried (Fig. 8) in an area that, apart from one single cremation burial, was totally devoid of human burials. Copper alloy studs, rings, a chain, beads and a lunate pendant survived from the collar. Dog burials are often directly associated with human burials; a dog and red deer, for example, were found in a pit in the eastern cemetery (Barber & Bowsher 2000, 19). The presence of this unique burial raises the question of whether it represents the burial of a treasured pet or an offering associated with the local beliefs of the community.

Figure 8. Dog burial

A well dating to the 2nd century was excavated in the same area as the dog burial. The well is contemporary with the cemetery and suggests that it was used as a ritual shaft rather than for water extraction. Excavations at Swan Street *c*.175m to the east of Lant Street, revealed 15 wells or shafts, with evidence to suggest that some were ritual in origin. 'Killed' pots and animal remains were found as offerings and one shaft included the remains of three disarticulated dog skeletons and a human skeleton which was articulated but incomplete due to the selective removal of certain elements (Beasley, forthcoming). At Lant Street, a dog skeleton and disarticulated human remains were found in ditch fills, analysis of which has yet to be undertaken.

Grave goods

Grave goods were recovered from twenty-three (26%) of the Lant Street burials. These included such jewellery as gold earrings, copper alloy bracelets, an iron ring and a number of various other personal items, as well as pottery and glass vessels. Hobnailed shoes were found in three (3%) of the burials. Block-lifting of one of the shoes has enabled an x-ray to taken revealing a broad rounded toe with a line of hobnails around the edge of the sole.

Figures 9, 10 and 11. Aryballos, amphorisk and bone inlay

The most impressive finds came from one of the chalk burials, that of a young adult female. This young woman was buried with a glass vessel on either side of her head and a group of objects, including a box and folding knife, at her feet. The glass vessels consisted of a late 2nd or 3rd century oil flask (*aryballos*) (Fig. 9) and a glass vessel mimicking the shape of an amphora (*amphorisk*) with a trailed glass decoration and now thought to been integral part of a larger vessel (Fig. 10) (John Shepherd pers comm.). Located at the feet were the remains of a casket of which the copper alloy fittings and carved bone inlay survived (Fig. 11). The inlay included the depiction of a female bust, set under a gabled pediment that formed the centrepiece. This style of decoration is seen on many tombstones in Britain (Major, 2005: 115). The folding knife, mentioned above, had an elaborate ivory handle carved into the form of a leopard and is probably of continental origin (Fig. 12). Also found with this group of objects was a copper alloy key, too large to be associated with the casket.

Figure 12. Folding knife

Demography

Of the eighty-nine individuals from Lant Street, fifty three were identified as adult (59%) and twenty two as juvenile (25%), the remaining fourteen (16%) were too poorly preserved for an age estimation. In contrast to America Street, where no juveniles below the age of six were present, the Lant Street assemblage included several young individuals. These include five individuals aged 3 to 6 years at death, one aged about 6 months, one aged about 9 months and a neonate.

Within the adult fraction of the assemblage, nearly half (43%) could not be assigned a sex due to poor preservation of the relevant skeletal elements. Of those that could be analysed, 22% were female, 25% were male and 10% were of intermediate sex. This gives a male to female ratio of 1.1:1, which when compared with both America Street and the eastern cemetery shows a slightly higher ratio of females to males.

In general there was very little identifiable pathology present within the assemblage. Of the pathologies observed, joint disease, fractures, including a contre-coup fracture to a tibia and fibula (the result of a blow to the side of the knee, a modern day comparison of this is often seen in pedestrians involved in road traffic accidents), infections such as periostitis and one case of possible osteomyelitis were included. The female buried with the casket and folding knife exhibited some unusual dental pathology including a talon cusp (a form of supernumerary cusp) on the lingual aspect of the left first incisor. The talon cusp had led to unusual wear patterns on both the mandibular and maxillary incisors, with the cusp causing both an over and under bite depending on the position of the cusp.

Cremations

Unfortunately neither of the cremation burials was very well preserved. One yielded only 22g of cremated bone from which no ageing or sexing data could be retrieved. From the second, it was possible to identify an adult individual but again no further ageing or sexing data was retrieved. A pottery vessel from one of the cremations was possibly of continental origin.

Conclusion

Prior to the excavations at America Street and Lant Street, the burials found within Southwark numbered 116 (Barber & Hall 2000, 104-7). Following these excavations, this number has tripled (to 374) and these discoveries enable us to broaden the picture of the population of this area, their beliefs and practices and enables comparisons to be made between both these sites and other Roman cemeteries within London.

Although post-excavation work is still continuing on both sites, a few basic comparisons can already be made. Although burial alignment is seen to vary between sites, the majority of burials were supine and extended, as would be expected. America Street had a high percentage of crouched and multiple burials, whilst Lant Street contained a higher percentage of prone and a larger proportion of female and infant burials. A number of grave goods are of possible continental origin reminding us of the wide movement of people and objects during this time. Cemeteries are an expression of ritual behaviour. However, what is important to note is the increasing evidence of an extensive ritual landscape around Lant Street, which has been illustrated by such sites as Swan Street (Beasley, forthcoming) and Tabard Square (Killock, forthcoming). The full potential of both of the sites discussed here will only be revealed when further research has been completed. What we do know, however, is that these results will help shed light on this less well known area of Roman London.

Acknowledgments

The authors would like to thank BABAO for the opportunity to present these summary results at both the 2005 conference and in these Proceedings.

Melissa Melikian would like to thank Schroders Property Investment Management Limited for funding the work. Specialist reports were prepared by Angela Wardle (accessioned finds), Sylvia Warman (animal bone), Matthew Conti (chalk analysis), Martin Evison (DNA), Robin Symonds (Roman pottery) and Michael Hammerson (coins). The illustrations were prepared by Jonathan Moller and Chris Adams.

Kathelen Sayer would like to thank Acorn Homes for generously funding the work at Lant Street. Thanks also go to Josephine Brown and Hayley Baxter for the drawings, Cheryl Blundy for the photographs and Hilary Major, Malcolm Lyne and Lisa Yeomans for their work on the small finds, pottery and animal bone respectively. Jim Leary and Frank Meddons are thanked for their comments and advice.

Literature Cited

Alcock J P (1980) Classical Religious Belief and Burial Practice in Roman Britain. Archaeological Journal 137: 50-85.

Allason-Jones L (1996) Roman jet in the Yorkshire Museum, York.

Barber B and Hall J (2000) Digging up the people of Roman London: Interpreting evidence from Roman London's Cemeteries. In I Hayes, H Sheldon and L Hanningan (eds.): London Under Ground: the archaeology of a city. Oxbow Books: Oxford: 102-120.

Barber B and Bowsher D (2000) The Eastern Cemetery of Roman London: Excavations 1983-1990. MoLAS Monograph Series 4.

Beasley M (Forthcoming) Excavations at the Old Sorting Office, Swan Street, Southwark.

Bentley D and Pritchard F (1982) The Roman cemetery at St Bartholomew's Hospital. Transactions of the London and Middlesex Archaeological Society 33: 134-72.

Clarke G (1979) The Roman cemetery at Lankhills, Winchester Studies 3: Pre-Roman and Roman Winchester: Part II.

Drummond-Murray J and Thompson P (2002) Settlement in Roman Southwark. MoLAS Monograph 12

Giorgi J (2000) The Plant Remains – a Summary. In MacKinder A: A Romano-British cemetery on Watling Street: Excavations at 165 Great Dover Street, Southwark, London. MoLAS Archaeological Study Series 4: 65-66.

Goodburn R (1978) Roman Britain in 1977. Britannia 9: 403-471.

Green MJ (1983) The Gods of Roman Britain. Shire, Aylesbury.

Hall J (1996) The Cemeteries of Roman London. In J Bird et al. (eds.): Interpreting Roman London – Papers in memory of Hugh Chapman. Oxbow Monograph 58: 57-84

Killock D (Forthcoming) Tabard Square assessment report. PCA report.

MacKinder A (2000) A Romano-British cemetery on Watling Street: Excavations at 165 Great Dover Street, Southwark, London. MoLAS Archaeological Study Series 4.

Major H (2006) Assessment of the small finds. In An assessment of an archaeological excavation at 52-56 Lant Street, London Borough of Southwark. Unpublished PCA report.

Melikian M (forthcoming) Excavations within the Roman 'Southern Cemetery'. Land adjacent to 1 America Street. AOC Archaeology Group Monograph.

Philpott R (1991) Burial practices in Roman Britain: A survey of grave treatment and furnishing, AD 43–410, BAR British Series 219. Oxford.

Roberts C and Cox M (2003) Health & Disease in Britain. Gloucestershire: Sutton Publishing.

Sayer K (Forthcoming) Excavations at Lant Street PCA Monograph Series.

Sheldon H (2000) Roman Southwark. In I Hayes, H Sheldon and L Hanningan (eds.): London Underground. Oxford: Oxbow Books: pp121-150.

Taylor A (2001) Burial Practise in Early England. Stroud: Tempus.

Taylor A (2003) Burial with the Romans. British Archaeology 69: 14-19.

Thomas C (2004) Life and death in London's East End: 2000 years at Spitalfields. MoLAS, London.

Toynbee JMC (1996) Death and Burial in the Roman World. London: Thames and Hudson.

Watson S (2003) An excavation in the western cemetery of Roman London. MoLAS Archaeology Studies Series 7.

Yule B (1988) Natural Topography of North Southwark. In Hinton P (ed.) Excavations in Southwark 1973-76 Lambeth 1973-9. Joint publication No.3. LAMAS, SAS: 13-17.

Yule B (2005) A prestigious Roman building complex on the Southwark waterfront. MoLAS Monograph 23.

The funerary behaviour and the social value of children in a proto-industrial urban population from London during the 18th and 19th centuries.

Funerary behaviour in a proto-industrial urban community

S.M. Bello* and L.T. Humphrey

Dept of Palaeontology
The Natural History Museum
Cromwell Road
London
SW7 5BD
UK
*e-mail address for correspondence: s.bello@nhm.ac.uk

Abstract

The skeletal and funerary remains recovered from the crypt of Christ Church Spitalfields provide a remarkable insight into burial practice in the 18th and 19th centuries. Differences in burial treatment directly or indirectly reflect the wealth and perceived status of the deceased. Data for age, sex and place of residence for the coffin plates sample from Christ Church (identified individuals known to have been interred within the crypt) is compared to comparable data from the burial register reflecting three decades of burials in both the cemetery and crypt. The coffin plate sample is characterised by a lower proportion of subadult burials, particularly those of females, and by a higher proportion of individuals who were resident in other parishes at the time of death. Differences in burial treatment between individuals within the crypt sample and variation in osteological preservation are explored.

Keywords: Burial practices, 18th and 19th centuries, subadults, London,

Introduction

This paper aims to document the funerary behaviour of a proto-industrial urban population from London during the 18th and 19th centuries. For this purpose, two sources of data have been analysed; the burial registers for Christ Church, Spitalfields (held at the London Metropolitan Archives), and the evidence of burial practice recovered during the excavation of Christ Church, Spitalfields. The comparative analysis of archival and archaeological sources aims to determine whether there was any selective recruitment of the population buried in the crypt of Christ Church according to the individuals' age and sex.

Christ Church, Spitalfields (London, U.K.) was consecrated in 1729 and its vaults were utilised as a repository for approximately 1,000 single or occasionally multiple interments between 1729 and 1857 (Reeve & Adams 1993). The crypt at Christ Church extends beneath the entire area of the building and includes several small vaults on the ground floor (Cox 1996). The funerary practices associated with the individual interments were thoroughly documented as part of the archaeological investigation of the crypt in 1984-6 (Reeve & Adams 1993). Legible coffin plates giving details of name and age were recovered from 383 coffins and the age at death, though not the name, is available for a further six skeletons (Molleson & Cox 1993). These individuals have been referred to as the named or coffin plate sample.

We suspected a bias in the constitution of the coffin plate sample, and preliminary analysis revealed two unexpected patterns. The mortality rate for individuals under the age of 15 years was only 20.5% in the coffin plate sample (Rousham & Humphrey 2002), whereas the Bills of Mortality reveal that 45-57% of recorded deaths were of juveniles in the overall London population during the18[th] and 19th centuries (Molleson & Cox 1993). Data for the coffin plate sample also revealed an imbalance in the numbers of male and female deaths during childhood, with 24% of males in the sample dying before the age of 15 years, compared with only 16.8% of females (Rousham & Humphrey 2002).

We will discuss here four possible causes for this bias: 1) taphonomic bias of the osteological sample with differential preservation of skeletal remains between demographic groups, 2) selective burial of individuals within the crypt according to demographic and/or social criteria, 3) bias caused by a relatively high proportion of non-resident adults within the coffin plate sample, and 4) differences in mortality patterns between the community represented by the coffin plate sample and the burial community as a whole. We will also explore whether a preferential burial in the crypt of individuals who held a higher social status within the community would reflect the socio-economic organization of this population with particular emphasis on the social value of children.

Materials and methods

Three different sources of information were examined for this study.

The first source was data derived from the burial registers for Christ Church Spitalfields, which are stored at the London Metropolitan Archives, which list all recorded intra and extramural burials. Each burial record gives the name and surname of the individual, their age at death (in days, months and years), the address where they resided (for most records). Some records are annotated with additional information, such as "son/daughter of ..." or "unknown person". The burial registers do not distinguish between those interred in the crypt and those buried in the cemetery surrounding the church, but those buried within the crypt represent a small percentage of the total number of burials. Records for all of the burials in three separate 10-year periods were transcribed into an Excel database. The completed database contains details of more than 15,000 burials.

The second source was the bibliographical data associated with individuals in the osteoarchaeological collection recovered from the crypt Christ Church Spitalfields. In many cases, surviving inscriptions on the coffin plates associated with skeletal remains revealed the name, age at death and date of death of the deceased, which allowed these individuals to be recognised in other historical records. Demographic analysis of the coffin plate sample is based on 380 individuals for whom age, sex and date of death are recorded.

Third, differences in burial practice among the coffin plate sample were examined. In particular three parameters were considered according to age: the type of coffin used for interment, the orientation of the coffins, and the distribution of burials within different areas of the crypt. For coffin type, the number of shells and the types of material used for construction were evaluated. The number of shells was scored as 1, 2 or 3, and the construction materials were classified as wood, lead, or wood and lead. The orientation of the coffin was considered both in terms of alignment (east-west or north-south) and angle (horizontal or vertical). The distribution of coffins within the crypt was evaluated in terms of the type of vault from which the coffin was recovered. Following construction of the church, the crypt was partitioned by internal walls in order to create more discrete areas for private burial or to ensure convenient access to all areas that were likely to be used (Reeve & Adams 1993). Specifically, three types of vault can be distinguished; private vaults, which were reserved primarily for the burial of members of a single family (there is evidence that these vaults had individual doors), public vaults, which were unreserved and included both related and unrelated individuals (family members may lie in spatial proximity to each other but this seems to occur when their deaths were temporally close), and redeposited vaults that were used for re-

stacking coffins removed from other areas of the crypt (Reeve & Adams 1993).

Results

Archival data

The number of deceased recorded in the burial registers for Christ Church, Spitalfields during each of the three ten-year periods examined is relatively constant (figure 1). The numbers of individuals identified as having lived within the Parish of Christ Church Spitalfields, in an adjacent parish, or in a more distant parish are given in Table 1. Adjacent parishes were defined as those bordering Christ Church Spitalfields. These are Bethnal Green, Mile End New Town Whitechapel, Algate, Bishopsgate and Shoreditch, as well as the extra parochial liberties of Norton Folgate and Old Artillery Ground. In some cases the address at the time of death was not recorded, and this includes all individuals buried between 1755 and 1759. In other cases the address was illegible or could not be located to parish, either because the street name occurred in several parishes or because it could not be identified on any map or because the information was insufficient (e.g. "found beneath the church"). Where the address is given as a hospital or workhouse it was not possible to identify the parish. Within the sample for which the address or parish was identified, 8,477 (77.6%) of the individuals buried at Christ Church resided within the parish, whereas 2,454 individuals (22.4%) resided in other parishes (Table 1).

The sex of the individual buried was determined from the first name for 14,773 of the 15,031 individuals represented in the database. Among these 14,733 individuals, 7,652 (51.8%) are female and 7,121 (42.8%) are male (Figure 1). A further 184 records did not give the name of the individual buried because the person was unknown or was a stillborn for whom the sex had not been recorded. In 74 cases it was uncertain whether the name was that of a male or female.

The age pyramids for the each of the three periods considered are similar and are all characterised by high percentages of deaths of subadult individuals, particularly those aged between 1 month and 1 year. Within the burial register sample, 46.1% of individuals have an age of death of below 20 years. Moreover, during each of the three periods, females are relatively under-represented in the subadult sample and over-represented in the adult sample compared to males. The difference in the proportion of males and females is statistically significant for subadults in the period 1750-1759 (χ^2=4.797, p=0.0285) and for adults in the periods 1750-1759 (χ^2=47.981, p<0.0001) and 1790-1799 (χ^2=14.867, p=0.0001; Table 2; Figure 2).

Table 1. Number of interments recorded in the burial registers for Christ Church, Spitalfields during each of the three 10-year periods considered.

Address in at time of death	Burial register sample		Coffin plate sample	
	No.	%	No.	%
Christ Church Spitalfields	8477	77.6	140	38.2
Adjacent parish or extra-parochial liberty	1863	17.0	122	33.2
More distant parish	591	5.4	105	28.6

Table 2. Number of interments recorded in the burial registers for Christ Church, Spitalfields according to the sex and age of individual.

	Burial registers						Coffin Plate Sample	
	1750-1759		1790-1799		1830-1839			
	female	male	female	male	female	male	female	male
stillborn	8	11	0	1	1	2	0	1
< 1 week	59	88	32	39	23	18	0	1
1week-1month	84	97	58	56	22	23	5	2
1month-1year	400	507	327	378	179	235	10	16
1-2 years	324	322	243	244	165	193	7	10
2-3 years	182	179	128	122	101	101	3	7
3-4 years	102	103	73	65	66	60	3	2
4-5 years	62	55	38	40	40	52	0	4
5-10 years	106	97	99	82	87	80	0	4
10-19 years	8	11	0	1	1	2	0	1
Totals	**1389**	**1534**	**1085**	**1101**	**757**	**835**	**38**	**50**
20-29 years	62	75	87	74	73	71	10	3
30-39 years	182	138	141	115	146	132	12	8
40-49 years	254	159	185	171	222	216	11	22
50-59 years	248	201	190	201	245	266	19	15
60-69 years	274	217	246	216	232	256	31	27
≥70 years	232	179	294	209	286	273	23	40
Totals	**1510**	**1079**	**1324**	**1092**	**1532**	**1407**	**145**	**147**

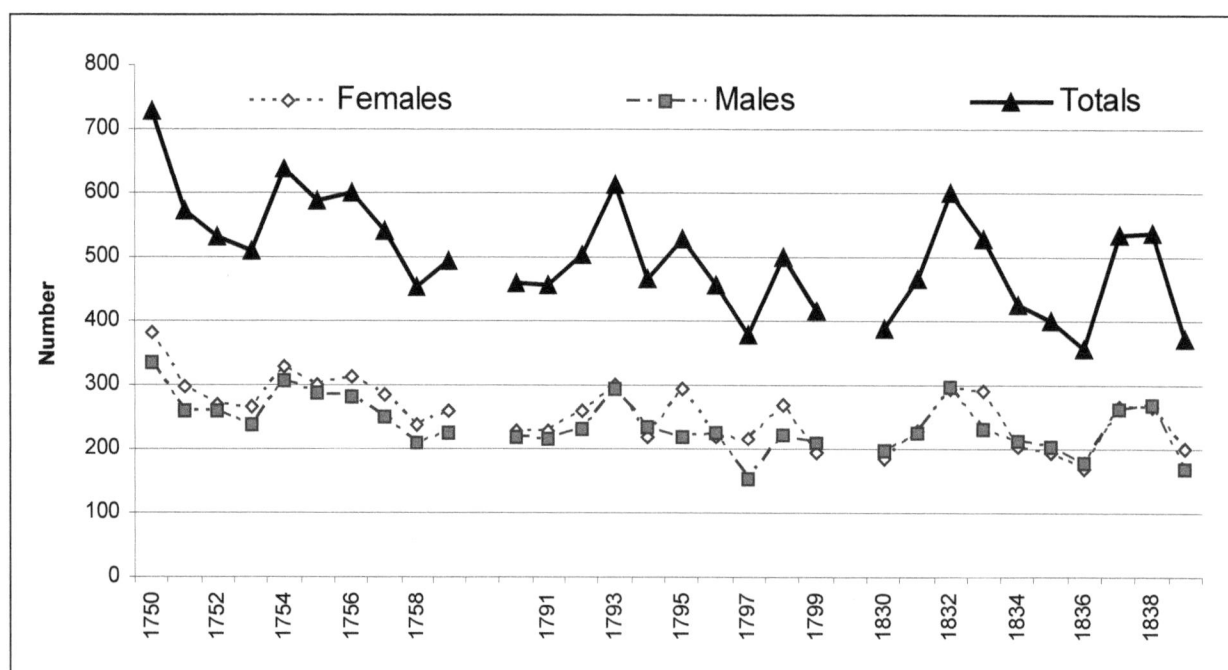

Figure 1. Number of interments recorded in the burial registers for Christ Church, Spitalfields during each of the three 10-year periods considered according to the sex of individual.

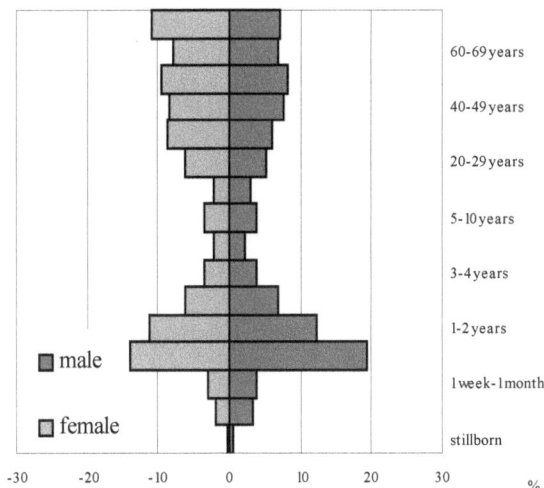

Figure 2a. 1750-1759

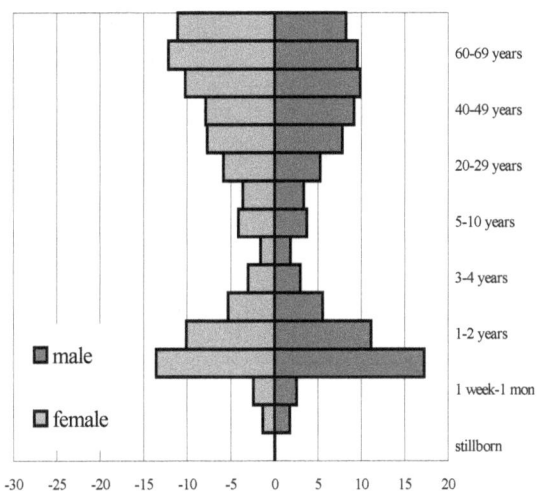

Figure 2b. 1790-1799

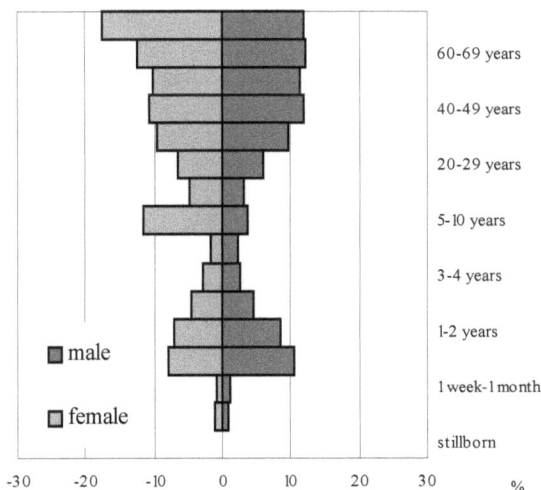

Figure 2c. 1830-1839

Figure 2. Age pyramids for all interments in Christ Church, Spitalfields during three ten-year periods. Females are shown on the left (in grey) and males on the right (in black).

Coffin plate sample

Within the coffin plate sample, 23.2% of individuals have an age of death of below 20 years. The under-representation of individuals younger than 1 year old in the coffin plate sample is particularly evident (10.3% of females and 10.2% of males). Within the total subadult sample, 43.2% are female and 56.82% are male, but this difference is not statistically significant (cf. table 2). The coffin plate sample is therefore characterised by an under-representation of subadult individuals (χ^2=75.426, p<0.0001) and, among subadults, by a noticeable but non-significant under-representation of females. Among adults there is an equal representation of females and males (Table 2, Figure 3).

Coffin characteristics among the Coffin plate sample

Considering the material of the shells and the number of shells, data were obtained for a limited number of 187 individuals. The majority of the bodies were buried in coffins built of three shells. No statistically significant differences were observed in the use of coffins according to the age of the individual (table 3). Coffins may have been constructed from wood only, from lead only or from wood and lead. The majority of the individuals were buried in coffins made from wood and lead (78.6 % of all the coffins). The distribution of individual adults and subadults in coffins made from just wood (16 subadults and 24 adults) or from wood and lead (31 subadults and 116 adults) is comparable (table 3).

Considering the position of the coffins, data were obtained for a total number of 361 individuals. In the crypt, the majority of coffins for both subadult and adult individuals were found in an orthodox position (east-west alignment), although the number of adult coffins in this position is significantly higher than those of subadults (cf. table 4). The practice of positioning the coffins vertically occurs more commonly among subadults than among adults (Table 4).

The distribution of coffins in the private, public and redeposited vaults was obtained for the whole sample of 380 individuals. Differences in the representation of the subadult and adult individuals between the different vaults do not appear to be influenced by age at death (Table 5). Very few children were recovered from the redeposited vaults, suggesting that the practice of removing coffins from their original place of interment was largely restricted to adults, contrary to potential expectation (the smaller coffins used for sub-adults would have been easier to transport than the bigger coffins used for adults).

Table 3. Type of coffins (number of shells) used according to the individual age and sex.

	Subadults	Adults	Chi-square
One shell	16 (34.0% of the coffins of subadult individuals)	24 (17.1% of the coffins of adult individuals)	3.632, p=0.05660
Two shells	6 (12.8%)	30 (21.4%)	1.194, p=0.2744
Three shells	25 (53.2%)	86 (61.4%)	0.259, p=0.6106

Table 4. Position of coffins in the crypt according to the individual age and sex.

	Subadults	Adults	Chi-square
Orthodox position (east-west alignment)	34 (40.5% of the subadult coffins)	192 (69.3% of the adult coffins)	5.853, p= 0.0155
Non-orthodox position (north-south alignment)	34 (40.5%)	77 (27.8%)	2.454, p= 0.1172
Vertical position	16 (19.0%)	8 (2.9%)	22.043, p= 0.0000

Table 5. Distribution of coffins in the private, public and redeposited vaults according to age and sex of the deceased.

	Subadults	Adults	Chi-square
Private vaults	15 (17.0% of the subadult coffins)	75 (25.8% of the adult coffins)	1.792, p= 0.1807
Public vaults	70 (79.5 %)	192 (65.1%)	1.059, p= 0.3034
Redeposited vaults	3 (3.4%)	25 (8.6%)	2.33, p= 0.1268

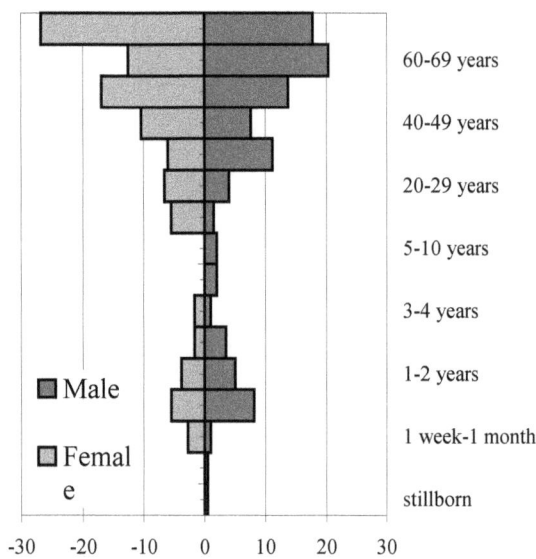

Figure 3. Age pyramid for the coffin plate sample

Discussion

The results of this analysis confirm that subadults in general and young females and infants in particular are underrepresented within the coffin plate sample relative to the total number of burials at Christ Church Spitalfields. Several different factors could contribute to this demographic discrepancy, including differential preservation, differential mortality within the section of the community represented by the coffin plate sample, and differential burial treatment according to demographic criteria.

Differential preservation of skeletal remains

Both in anthropological and zooarchaeological studies, there is a general tendency to assume that the remains of subadults are less well preserved than those of adults. Inherent differences in size, shape and density of different skeletal elements are responsible for variable rates of decay, not only in the chemical breakdown of bone, but also in the role played by extrinsic factors in decomposition (Henderson 1987). Experimental work (Bouchud 1977; Lambert et al. 1985; Von Endt & Ortner 1984) has shown that the rates of decay are inversely proportional to bone size. A poorer state of preservation and representation of the smaller bones of the skeletons, both for subadult and adult individuals has been observed consistently in other human osteological assemblages (Bello 2001; Bello et al. 2002, 2006: figure 4). Subadult bones are typically smaller than those of adults, which increases their vulnerability to decay and increases the chances that they will be overlooked during excavation. Moreover, since the relative volume of bones is related to the individual's age, it is probable that the bones of the younger infants would be even more affected than the bones of elder subadults.

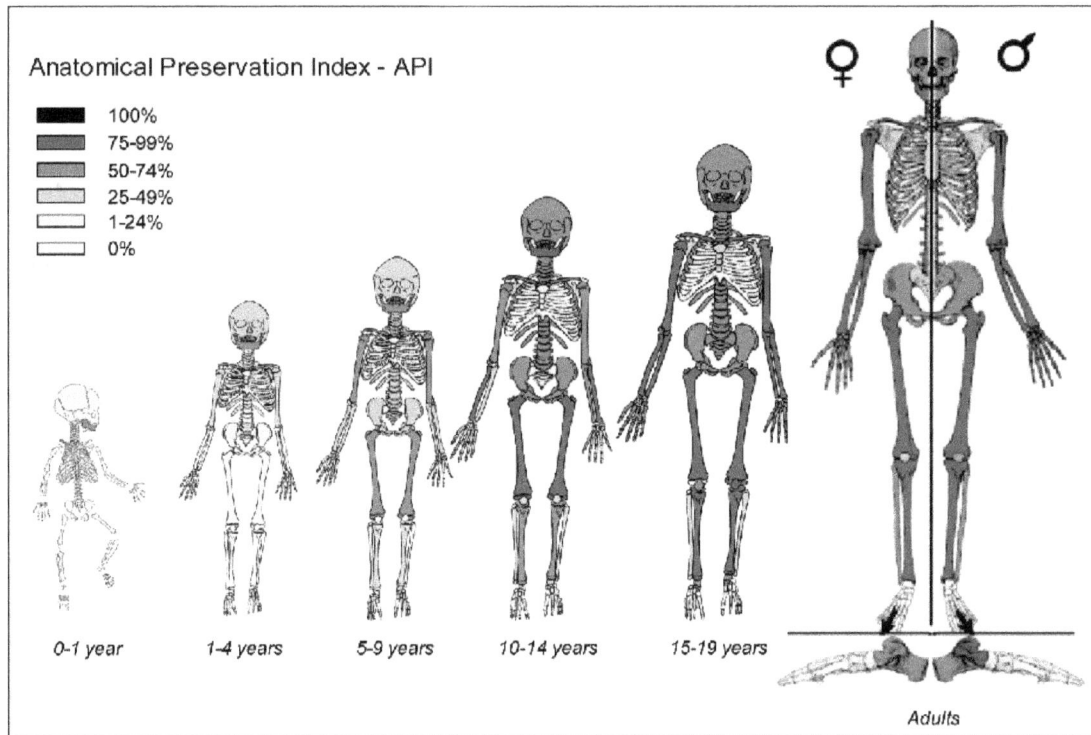

Figure 4. State of preservation of Délos sample (plague site, 1722, Martigues, France). Darker colour indicates a better state of preservation.

The state of preservation of the coffin plate sample from Christ Church, Spitalfields has been discussed by Bello et al. (2006). The analysis revealed that subadult individuals were generally less well preserved than adults, and particularly within the youngest age classes (0-4 years). In females, a poor state of preservation was observed for both the age classes of less than 1 year and 1-4 years. In males, the poorer state of preservation was only observed for individuals aged less than 1 year (Bello et al. 2006). This pattern of differential preservation could contribute to the under-representation of subadult individuals, particularly subadult females within the sample recovered. We think, however, that the substantial under-representation of subadults observed at this site cannot be exclusively explained by a poorer preservation of immature remains, and that other reasons should be considered.

Preferential treatment of particular demographic groups

Funeral and burial practices can select a portion of the population according to biological (sex, age or family relations) and/or social criteria (Castex et al. 1996; Crubézy et al. 1990; Dedet et al. 1991; Duday & Masset 1987; Gnoli & Vernant 1982; Masset 1986; Masset 1997; Sansilbano-Collilieux 1990; Tillier & Duday 1990). The exclusion of all or a part of demographic subset of the community, as well as the use of different burial structures according to the biological and/or social status of the deceased can have a substantial impact on the demographic interpretation of an osteological sample. In cases where such biases can be reliably inferred, they

can provide an unusual insight into the socio-economic organization of a community.

At Christ Church, as at any other parish church, those who died in the parish had a right to be buried in the graveyard. Depending on their financial standing and personal reasons, they might choose to be buried not in the churchyard, but within the walls of the church itself.

The security provided by the walls of a crypt was an attractive alternative to the overcrowded cemetery, and protected against the risk of being disturbed or removed by resurrection-men who, between 1750 and 1830, supplied hospitals with bodies for dissection and study (Hewer 1954).

King et al. (2005) suggested subadults are under-represented in the coffin plate sample because they did not have the same social status as adults, and that the higher representation of male children in the coffin plate sample from Spitalfields is an indication of higher male status rather than higher male mortality. Interestingly, within the crypt, subadult and adult individuals were interred according to different criteria. Children were typically buried in simpler coffins with only 1 or 2 shells and were less likely to be placed in the private vaults.

Burial in private areas of the crypt and in more sophisticated coffins would have added to the funeral cost and these options would have been available only to wealthier and more privileged individuals.

The evidence that a higher number of subadult coffins were positioned vertically than those of adults might imply that the coffins of children were treated with less regard than those of adults, but this situation may have arisen simply because the subadult coffins could be more easily fitted into small remaining spaces in order to make way for more efficient horizontal stacking of the larger adult coffins. Ultimately space constraints rather than social status may have been the most critical factor controlling the final positioning of the coffins at Christ Church. The shape and architecture of the crypt, the size of the disposal area, and the constraints imposed on disposal by difficulties of access were major factors in the maintenance, rearrangements and deliberate disturbance of the coffins. Interments placed on a north-south alignment, contrary to orthodox practice, or placed vertically, strongly indicate the over-riding concern with maximising use of space (Reeve & Adams 1993).

Burial of non-parishioners within the crypt

Another aspect of burial behaviour is choice of burial location. Cox (1996) noted that among the coffin plate sample, individuals who were resident in other parishes at the time of death were predominantly adults. The percentage of individuals resident in other parishes at the time of death is substantially higher (64.7%) in the coffin plate sample than in the total burial sample, and this difference is more marked for adults (64% non-resident) than for subadults (54.8% non-resident). This in turn suggests that the demographic profile of the coffin plate sample could be biased by the inclusion of a relatively high proportion of non-resident adults. In fact, the effect of this bias is relatively minor as adults accounted for 72.8% (102/140) of the resident coffin plate sample and 79.7% (181/227) of the non-resident coffin plate sample. Cox (1996) suggests that adults may have chosen to be buried close to other family members and in the parish where they were baptised or raised, whereas children were more likely to be buried in the most convenient location. It is also possible that there was a general tendency for wealthy individuals to be buried in their parish of birth and that children were less likely to have moved away from that place at the time of death.

Socioeconomic factors

Part of the difference in the proportion of subadult deaths between the coffin plate sample and the burial sample as a whole may be due to a genuine difference in subadult mortality between the section of the community who selected burial in the crypt and the wider Spitalfields community. Those who could afford a crypt burial are likely to have been socially and economically advantaged in other respects and would have had access to less crowded housing and better nutrition. These families may also have been more influenced by changing fashions relating to infant nutrition, which would not always have been beneficial to infant and child health (Cox 1996). We have not been able to test explicitly the hypothesis that socioeconomic advantages contributed to a lower subadult mortality within the families represented in the coffin plate sample.

Conclusion

Comparison of the age pyramids obtained for the coffin plate sample and the burial register for Christ Church Spitalfields reveals that the main difference concerns the representation of the subadults: only 23.2% of the coffin plate sample is composed of subadults, while the burial registers recorded that 46.1% of the burials were of subadults. Several different parameters that may have contributed to this discrepancy are identified. The smaller size of subadults and simpler coffin types afforded to younger individuals may have resulted in poorer preservation of immature remains within the crypt and therefore a lower representation of subadults in the recovered sample than in the original crypt sample. The coffin plate sample is characterised by a substantially higher proportion of individuals who were resident in other parishes at the time of death than the burial sample as a whole. The percentage of non-resident individuals is higher among adults than among subadults, and we estimate that the effect of this bias accounts for about 4% of the apparent difference in subadult mortality between the coffin plate sample and the total burial sample. Childhood mortality may have been lower within families represented by the coffin plate sample as a result of other advantages associated with their relatively high social status and prosperity. Nevertheless it is likely that a significant part of the difference is due to the tendency to exclude children from a prestigious and more expensive burial in a crypt and that this in turn reflects the fact that children occupied a lower social status than adults in this proto-industrial community. At the same time, this result could also suggest that the children buried inside the crypt were somehow "special" compared to those in the rest of the community.

The apparent under-representation of young girls in the coffin plate sample (43.2% of individuals under 20 years) was not statistically significant, and notably, the percentage of subadult females present in the total burial sample was also less than half (48.2%). Results of this study do not therefore suggest that female children were preferentially excluded from a burial in the crypt. Differences in preservation between male and female children, particularly those aged between 0 and 4 years, may also have contributed to the apparent imbalance in the numbers of male and female subadults in the coffin plate sample. Within the coffin plate sample there is no evidence for differences in the type of coffin used for interment, the orientation of the coffins, and the distribution of burials within different areas of the crypt between males and females.

Acknowledgements

We would like to thank Theya Molleson, Peter Andrews, Tania King, and Emily Rousham for fruitful and helpful discussions, and two anonymous reviewers for their helpful comments.

Literature Cited

Bello S, Thomann A, Signoli M, Dutour O and Andrews P (2006) Age and sex bias in the reconstruction of past population structures. American Journal of Physical Anthropology 129: 24-38.

Bello S, Signoli M, Rabino-Massa E and Dutour O (2002) Les processus de conservation différentielle du squelette des individus immatures. Implications sur les reconstitutions paléodémographiques. Bulletins et Mémoires de la Société d'Anthropologie de Paris 14: 245-262.

Bello S (2001) Taphonomie des restes osseux humains. Effet des processus de conservation du squelette sur les paramètres anthropologiques. PhD dissertation: Università degli Studi di Firenze and Université de la Méditerranée.

Bouchud J (1977) Etude de la conservation différentielle des os et des dents. In Approche écologique de l'homme fossile. Travaux du groupe – Ouest de l'Europe de la Commission Internationale de l'INQUA Palecology of Early Man (1973-1977), In H Laville and J Renault-Miskovsky (eds.) : Université Pierre et Marie Curie: Paris : 69-73.

Castex D, Courtaud P, Sellier P, Duday H and Bruzek J (eds.) (1996) Les ensembles funéraires: du terrain à l'interprétation. Bulletins et Mémoires de la Société d'Anthropologie de Paris 8: 527.

Cox M (1996) Life and Death in Spitalfields 1700 to 1850. Council for British Archaeology: York.

Crubézy E, Duday H, Sellier P and Tillier AM (eds.) (1990) Anthropologie et Archéologie dialogue sur les ensembles funéraires. Bulletins et Mémoires de la Société d'Anthropologie de Paris 2: 226.

Dedet B, Duday H and Tillier AM (1991) Inhumation de fœtus, nouveau-nés et nourrissons dans les habitats protohistoriques du Languedoc: l'exemple de Gailhan (Gard). Gallia 48: 59-108.

Duday H and Masset C (eds.) (1987) Anthropologie physique et archéologie – Méthodes d'étude des sépultures. Edition du CNRS: Paris.

Gnoli G and Vernant JP (1982) La mort, les morts dans les sociétés anciennes. Cambridge University Press: Cambridge.

Henderson J (1987) Factors determining the state of preservation of human remains. In A Boddington, AN Garland and RC Janaway (eds.): Approaches to archaeology and forensic science, Manchester: Manchester University Press: 43-54.

Hewer RH (1954) The sack-em-up men. St Bartholemew's Hospital Journal June.

Lambert JB, Simpson SV, Weiner JG and Buikstra JE (1985) Induced metal-iron exchange in excavated human bone. Journal of Archaeological Science 12: 85-92.

Masset C (1986) Le recrutement d'un ensemble funeraire. In H Duday and C Masset (eds.): Anthropologie physique et Archéologie. Paris: Edition CNRS : 108-126.

Masset C (1997) Les dolmens. Sociétés néolithiques et pratiques funéraires. Paris: Edition Errance.

Molleson T and Cox M (1993) The Spitalfields Project. Volume 2 – The Anthropology. The Middling Sort. CBA Research Report 86. Council for British Archaeology: York.

King T, Humphrey LT and Hillson S (2005) Linear enamel hypoplasias as indicators of systemic physiological stress: evidence from two known age-at-death and sex populations from Post-Medieval London. American Journal of Physical Anthropology 128: 547-549.

Reeve J and Adams M (1993) The Spitalfields Project. Volume 1 – The Archaeology. Across the Styx. CBA Research Report 85. Council for British Archaeology: York.

Rousham EK and Humphrey LT (2002) The dynamics of child survivorship. In: H MacBeth and P Collinson (eds.): Human Population Dynamics. Cambridge University Press : 124-140.

Sansilbano-Collilieux M (1990) Les caractère discrets et le « recrutement » de deux nécropoles du Haut Moyen Âge à Poitiers. Bulletins et Mémoires de la Société d'Anthropologie de Paris 2: 179-184.

Tillier AM and Duday H (1990) Les enfants morts en période périnatale. Bulletins et Mémoires de la Société d'Anthropologie de Paris 2: 89-98.

Von Endt DW and Ortner DJ (1984) Experimental effects of bone size and temperature on bone diagenesis. Journal of Archaeological Sciences 11: 247-253.

The 'French Disease': Syphilis and the burial ground of St Pancras

Natasha Powers[1]* and Phillip Emery[2]

[1] Human Osteologist
Museum of London Specialist Services
Mortimer Wheeler House
46 Eagle Wharf Road
London N1 7ED
UK

[2] Principal Archaeologist
Gifford
Pentagon House
52-54 Southwark Street
London
SE1 1UN UK

*e-mail address for correspondence: npowers@museumoflondon.org.uk

Abstract

During the eighteenth century, the "French disease", "French Pox" or the "Great Pox" was commonplace. Essentially an urban disease, syphilis was a recognised cause of death. The brothels of mid-eighteenth century London sold condoms to clients as a preventative for infections acquired from prostitutes. Syphilitics were banned from general hospitals. The period of the St. Pancras burial ground extension (1793-1854) marks a time of change in attitudes between the rakish Georgian and his more prudish Victorian counterpart. Treponemal infections were the most prevalent of the specific infections in the St. Pancras assemblage. There were four adult individuals for whom sufficient skeletal pathology was present to ensure a definite diagnosis of venereal syphilis and a further eight were probable victims of syphilis. This included several named individuals including two men with particularly florid skeletal changes. This poster investigates the demographic distribution in the St. Pancras assemblage with reference to contemporary data.

Keywords: Treponematosis, venereal syphilis, 'Pox', 'French Disease', mercury, named individuals

Introduction

The St. Pancras assemblage, excavated in advance of construction of the new London terminus for the Channel Tunnel Rail Link, provides a 'snapshot' into the health of Londoners in the late 18th and early 19th centuries; a time of population increase, industrialisation and a subsequent rise in the volume of urban poor. A team assembled by Gifford (Archaeology) carried out the archaeological investigations. The contrasting early, wealthy burials and later workhouse inhabitants emphasises a socially divided city (Roberts & Cox 2003). Eighty-two percent of the dated skeletal sample came from 1793-1812 and a substantial proportion of French *émigrés* were present (Emery, this volume). Staff from Pre-Construct Archaeology Ltd. and the Museum of London Archaeology Service undertook osteological analysis of 715 individuals: 532 adults (including 219 females and 231 males) and 183 sub-adults. The majority of the sub-adults (141: 77%) were aged below 5 years when they died.

The population of London came into regular contact with numerous infectious diseases in everyday life: bathing was not considered essential and houses and their inhabitants were parasite infested (Roberts & Cox 2003: 297). Many of the big killers of the time, the febrile and diarrhoeal diseases, are acute conditions and consequently are osteologically invisible. Treponemal infections (a group of diseases which include venereal syphilis) were the most prevalent of the specific infections seen in the skeletal assemblage.

The 'French Disease' in London

During the 18th century, the "French disease", "French Pox" or "Great Pox" as syphilis was known, was commonplace. Essentially an urban disease linked to the mobile armed forces and sailors, the "French Pox" was a recognised cause of death. Bills of Mortality indicate that between 1790 and the early 19th century, the death rate gradually declined from 0.3% of total deaths to 0.1% (Lane 2001: 153; Roberts & Cox 2003: 242).

The brothels of mid-18th century London sold condoms to clients, as a prophylaxis for infections acquired from prostitutes and were even recommended by that most famous lothario, Casanova. However, association with prostitution meant condom use was not entirely socially acceptable (Lane 2001: 37). Although syphilitics were banned from general hospitals, in the 1740s the Lock Hospital in Grosvenor Square was established to care for affected Londoners. The hospital boasted that of the 695 patients admitted in the first two years, 646 left cured, and only around 20 died (Lane 2001: 153). The period during which the St. Pancras burial ground was in use marks a time of change in attitudes between the rakish Georgian and the more prudish Victorian counterpart. By the late 18th century, an increasingly moralistic view was leading to stigmatisation of the victims (Roberts & Cox 2003: 341).

Skeletal manifestations of infection

Venereal syphilis, caused by the spirochete *Treponema pallidum pallidum*, is characterised by three stages: a primary lesion at the site of infection, a secondary skin rash, and tertiary changes affecting the skin, skeleton, central nervous and cardiovascular systems. Only in this latter stage is it lethal, with skeletal lesions appearing 2 to 10 years after the initial infection (Aufderheide & Rodriguez-Martin 1998).

Venereal Syphilis at the St. Pancras burial ground

There were four adult individuals for whom sufficient skeletal lesions were present to ensure a definite diagnosis of venereal syphilis and a further eight with probable syphilis (gummatous osteitis typical of the infection, but no observable cranial pathology). No cases of congenital syphilis were present in the sub-adult population. Twenty-three-year-old Mary Ann Shillito (died 1822), and Thomas Beck, an inmate of the nearby workhouse (died 1797) were two of those affected.

A further 14 individuals had bony changes described as *possibly* syphilitic. One individual, coal merchant Pierre Jossaume, was married to Jeanne Jossaume, who showed no osseous indications of the condition. Her Death Certificate stated cause as 'old age' (she was 84), though the possibility of the attending doctor exercising his discretion should not be discounted.

Demographic data for all probable syphilis cases is shown in Figure 1. The apparent peak in the 26-45 years old males is emphasised when all possible treponematoses are pooled (*Fig 2*). The crude prevalence rate for *probable* syphilis was 1.7% (12/715). If all *possible* cases were included this rises to 3.6% (26/715). The *probable* syphilitics represent 2.6% of adult males (6

of 231) and 2.6% of females (5 of 219), compared with 0.21-3.7% for other sites of the period from England (Roberts & Cox 2003: 341).

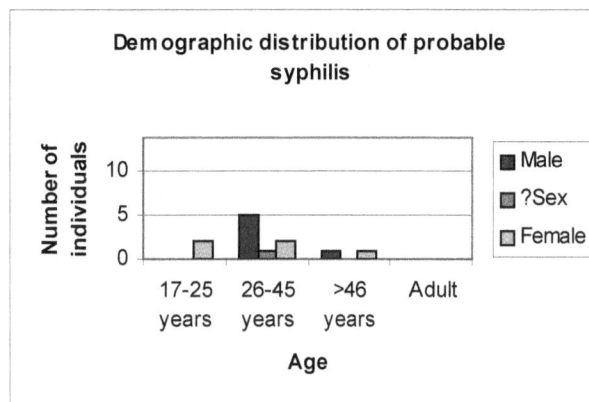

Figure 1. Demographic distribution of all definite and probable venereal syphilis cases

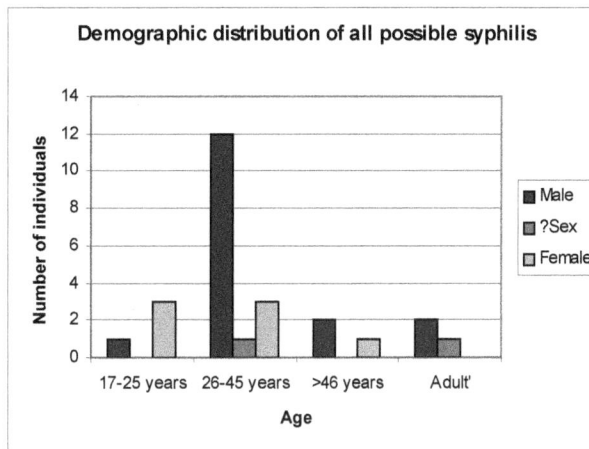

Figure 2. Demographic distribution of all possible Treponematosis cases (including sclerosing osteitis)

Named individuals

Two individuals with particularly florid skeletal changes, had biographic information: Mr Francis Coster (died 1798 aged 30) and a Thos[?] suggesting the latter was also male, despite the inconclusive osteological results.

Mr. Coster had advanced cranial changes: *caries sicca* with sclerosing osteitis associated with periods of healing and remodelling (Fig. 3). There was a proliferation of gummatous lesions, some penetrating both tables of the cranial vault. Gross, unilateral changes were present in the left rhino-maxillary area, the palatine process, incisive fossa and canine eminence all absent and a communication between the mouth and nasal aperture (Fig 4).

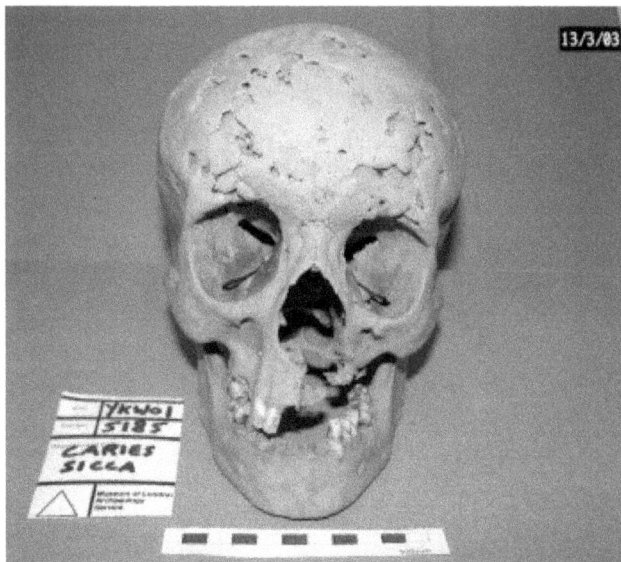

Figure 3. Mr. Francis Coster, showing cranial and facial changes as the result of venereal syphilis

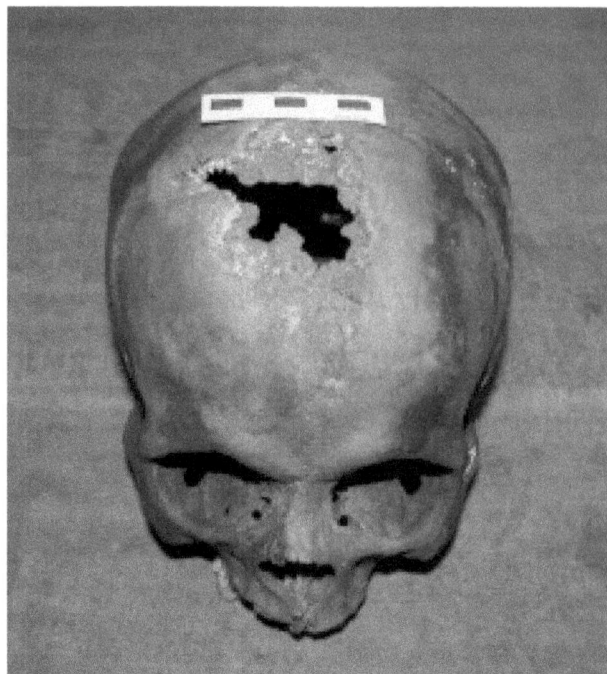

Figure 5. Destructive lesion in the superior frontal of Thos[?], note also the healed, stellate lesions in the inferior frontal

on the inferior part of the frontal. This indicates one episode of attack followed by a hiatus with some healing and latterly a more destructive episode, active at the time of death

The treatment of syphilitics

Cures and patent medicines for venereal diseases were widely advertised, and indeed satirised, in the 18th century (Lane 2001: 151). Paracelsus advocated the use of mercury as a treatment in the 16th century. Calomel, or mercurius dolcis (Hg_2Cl_2) was historically used to treat serious skin complaints and was adopted as a treatment for syphilis (Porter 1997:175). Mercury was considered a cure-all though its dangers were recognised (Lane 2001: 152). Unfortunately for the patients, the side effects of the treatment could be worse than the illness. From increased salivation and gastro-intestinal complaints to death: gum ulceration, tooth loss, loss of bone density and necrosis of the palate and alveolar bone were common. The high mortality rate of the cure was, in part, disguised by that of the disease itself (Porter 1997). It is interesting to speculate whether treatment altered the bony expressions of the disease, either directly or by increasing survivorship, given that the cranial vault changes seen at this site are so florid.

Discussion and Conclusions

Clinical cases peak between the ages of 20-24 years with a significant male bias (Aufderheide & Rodriguez-Martin 1998: 158). This fits the pattern seen at St Pancras.

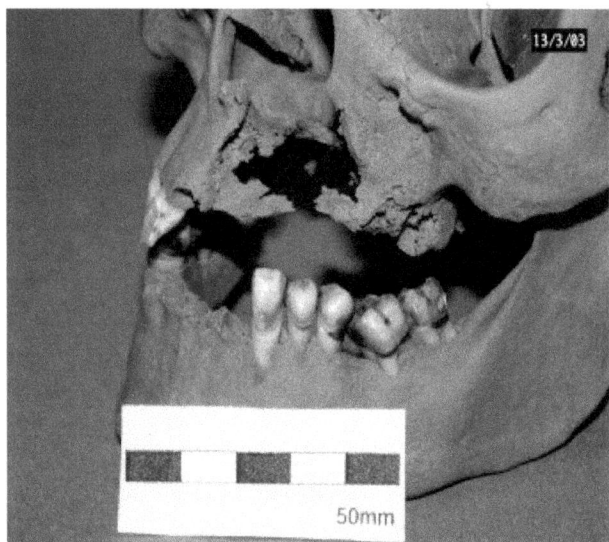

Figure 4. Rhino-maxillary destruction and ante mortem tooth loss

The teeth of the right maxilla had all been lost before death. Sclerosing osteitis with gummatous foci affected the left humerus and acromio-clavicular articulation. Sores, inflammation and reddened skin in the limb and pus from the affected mucous membranes draining through the communicating nose and mouth would have been symptomatic of this disease.

The mature adult identified only as Thos[?] had osteitis in the ulnae, femora and tibiae and a large, aggressively destructive lesion on the superior frontal bone (Fig 5). This sharp-edged lesion was surrounded by disorganised new bone growth. Healed stellate lesions could be seen

Elsewhere a single adult female from Red Cross Way had cranial changes resulting from venereal syphilis (Brickley & Miles 1999) and a further adult male with the condition was noted at Kingston-upon-Thames (Start & Kirk 1998). Christ Church Spitalfields had only two cases of venereal syphilis and Waldron (*in* Molleson & Cox 1993: 84) suggested this might indicate family stability within this population, acting as a significant limiting factor for sexually transmitted infections.

The crude prevalence rate for probable syphilis at St. Pancras is slightly higher than those so far seen for post-medieval London but within the expected range for the period. The demographic distribution appears to concur with that shown by clinical studies. The health and mortality risks to the population of Somers Town were many and varied, and dependent on subtleties of social status and behaviour that cannot be determined from the osteological data alone. However, if Waldron's suggestion to explain the dearth of cases at Christ Church is indeed true, this suggests a marked contrast in social behaviour between the two, roughly contemporary populations, despite each having a large component of French migrants. Perhaps, this demonstrates paleopathologically the different social spheres and religious backgrounds of the two groups.

Acknowledgements

The archaeological investigation at St Pancras burial ground formed part of an extensive programme of works along the route of the Channel Tunnel Rail Link, funded by London and Continental Railways. The authors also wish to thank all those involved in the project particularly Helen Glass and those involved in the osteological recording of the human remains and the compilation of data for analysis: James Langthorne, Hanne Rendall-Wooldridge, Kathelen Sayer, Don Walker, Bill White, Kevin Wooldridge (Fieldwork Director) and Tiziana Vitali.

Literature Cited

Aufderheide AC and Rodríguez-Martín C (1998) The Cambridge Encyclopaedia of Human Palaeopathology. Cambridge University Press: Cambridge.

Brickley M and Miles A (1999) The Cross Bones Burial Ground Redcross Way Southwark London. Archaeological Excavations (1991-1998) for the London Underground Limited Jubilee Line Extension Project. MoLAS Monograph 3 (MoLAS: London).

Lane J (2001) A Social History of Medicine: health, healing and disease in England 1750-1950. Routledge: London.

Molleson T and Cox M (1993) The Spitalfields Project. Volume 2: The Anthropology - The Middling Sort CBA Research Report No 86: York.

Porter R (1997) The Greatest Benefit to Mankind: a medical history of humanity from antiquity to the present. Harper Collins Publishers: London.

Roberts CA and Cox M (2003) Health and Disease in Britain: from prehistory to the present day. Sutton Publishing Ltd: Stroud.

Start H and Kirk L (1998) 'The bodies of Friends'- the osteological analysis of a Quaker burial ground. In Cox M (ed.): Grave Concerns: death and burial in England 1700-1850. CBA Research report 113: York: 167-178.

Industrial accident or deliberate amputation?
Three case studies from a Victorian population in Wolverhampton, West Midlands

Three cases of amputation from Wolverhampton

Paola Ponce[1], Iraia Arabaolaza and Anthea Boylston[2]

[1]Department of Archaeology
Durham University
South Road
Durham
DH1 3LE

[2]Biological Anthropology Research Centre
Department of Archaeological Sciences
University of Bradford
Bradford
BD7 1DP
e-mail address for correspondence: p.v.ponce@durham.ac.uk

Abstract

Three cases of amputation were found in an industrial population from the overflow cemetery of St. Peter's Collegiate Church in Wolverhampton (West Midlands). They were among 150 burials excavated from the cemetery, which was used for burial from AD 1819-1850 (Adams & Coll 2006). The amputations had been performed on the upper limb of a female (HB53) and lower limbs of two males (HB86) and (HB129). Two were surgical amputations and one appeared to be the direct result of an industrial accident. The only other amputations recovered to date from postmedieval contexts are an individual from Masshouse Circus in Birmingham (Brickley et al. in press) and another from St Marylebone churchyard in London (Miles et al. in prep), although the situation is changing rapidly with the excavation of more 19th century sites in London. The three cases from Wolverhampton provide examples of survival following traumatic surgical intervention at a time when anaesthesia and prevention of sepsis were in their infancy. This paper aims to place them in their historical and archaeological context.

Keywords: amputation, surgery, industrial, Wolverhampton, Victorian period, 19th century

Introduction

Amputation was the most common surgical procedure undertaken at the beginning of the 19th century, to judge by documentary sources (Porter 1997). According to Victorian surgeons, the pathological conditions necessitating the procedure were severe compound fractures, weapon injuries and other lacerated and contused wounds, gangrene, congenital deformities, osteomyelitis and various kinds of tumour (Duckworth 1980; Haeger 1988). Tuberculous joints also caused the affected limb to be amputated on occasion (Porter 1997).

In the first half of the 19th century, speed was the key to performing a successful amputation owing to the lack of adequate anaesthesia (Witkin 1997). Post-operative infection often began in an operating theatre that was sometimes full of the bystanders watching the operation as if it was a public spectacle. Surgeons wore dirty coats and used instruments that, although they may have been washed, were still covered with bacteria. The infection was also spread from patient to patient by using the same sponge to clean wounds (Wangensteen & Wangensteen, 1978).

By 1850 amputation techniques had begun to advance quickly, mainly due to two military surgeons, Baron Dominique Larrey and George Guthrie (Kostuik 1981). The surgical methods used to carry out an operation consisted of "circular incisions" or "flap operations" (Witkin 1997). The former was also called "tour de maître" and rapidly gained in popularity with surgeons. It consisted of one single circular stroke to divide skin and muscles. The skin was dissected back about five centimetres and rolled up to expose a large segment of bone. Following this, the bone was sawn and smoothed off using a rasping instrument. Joining together the muscles and the skin made the final stump, which was held in place with adhesive straps (Bell 1821). Flap operations followed the same technique but cut skin and muscles obliquely outwards leaving a flap of skin and muscles.

Regardless of amputation technique, the likelihood of survival was low, with a mortality rate among amputees of up to 60% (Porter 1997). However, after the discovery of anaesthesia, with chloroform in the 1860s, and the introduction of antisepsis, with the first carbolic acid spray, the number of fatalities fell (Roberts & Cox 2003). These varied from surgeon to surgeon and were dependent on their cleanliness procedure, wound care and initial contamination (Kostuik 1981; Wangensteen & Wangensteen 1967).

Despite their common occurrence in the historical record, cases of amputation in human skeletal remains at all periods are much scarcer. Mays (1996) reviewed the scanty archaeological evidence in the Old World and Merbs (1989) described occasional cases in the New World and in an Egyptian dating to the Ptolemaic period. Roberts and Manchester (2005:123) illustrated the amputation of the right foot of an individual with leprosy from Chichester. Hence, the three cases described in this report present a unique opportunity to examine in detail the hazards of life for an urban population from the heart of the Black Country.

Materials and Methods

An excavation was conducted by Birmingham University Field Archaeology Unit between October 2001 and January 2002 on the northern part of St. Peter's Collegiate Church overflow cemetery (Adams & Coll 2006). The project was undertaken as part of the planning conditions for an extension to the Harrison Learning Centre, University of Wolverhampton (Figure 1). A total of 152 graves were excavated, revealing 150 burials from an urban cemetery dating to the first half of the 19th century. Most of the individuals were probably buried in coffins, as suggested by the remains of wood and nails. Seven brick vaults were also found; six of which had been emptied during earlier graveyard clearance. As well as this previous vault clearance, some building works had taken place at the site, producing truncated graves. Only the south-eastern part of the site was not subject to prior clearance and consequently remained undisturbed (Neilson & Coates 2002).

As is normal for a Christian burial, the burials were aligned east-west, with the head at the western end of the grave. They were usually in a supine position, although the disturbance of the site by later buildings may have altered the location of some bones. In most cases, the position of different skeletal elements seemed to be consistent with the natural process of decay. A total of 96 adults were recovered; the other 54 burials were those of infants and children. Amputation was observed in three of the adults.

The techniques employed for age estimation were; changes in the morphology of the pubic symphysis (Brooks & Suchey 1990), the auricular surface (Lovejoy et al. 1985), closure of the vault and lateral-anterior cranial sutures (Meindl & Lovejoy 1985) and morphological changes in the sternal rib ends (İşcan et al. 1984; 1985). In some instances, the ossification of the thyroid cartilage (Krogman & İşcan 1986:128) was used to assess the age category, which was particularly useful for determining the age of the oldest individuals.

Figure 1a & 1b: Location of the excavation in the Harrison Learning Centre, part of Wolverhampton University.

Sex assessment was performed by studying the morphological characteristics of the cranium, mandible and pelvic region (Bass 1987; Buikstra & Ubelaker 1994; Ubelaker 1989) along with standard metrical assessment (Stewart 1979). Stature was calculated, where possible, using the formulae developed by Trotter (1970).

Each skeleton was observed macroscopically for evidence of pathology and, where necessary, X-rays were taken in order to assist with the diagnosis. Among the pathological cases identified, there were examples of healed trauma, particularly to the nasal and hand bones, and perimortal trauma. Infectious diseases included tuberculosis and syphilis.

Results

The first case (HB86) was a middle adult male (approximately 36-45 years of age). His stature was calculated as 1.74 m ± 3.27 cm. The amputation was carried out on the distal end of the left tibia and fibula transecting the shafts of both bones (Figure 2).

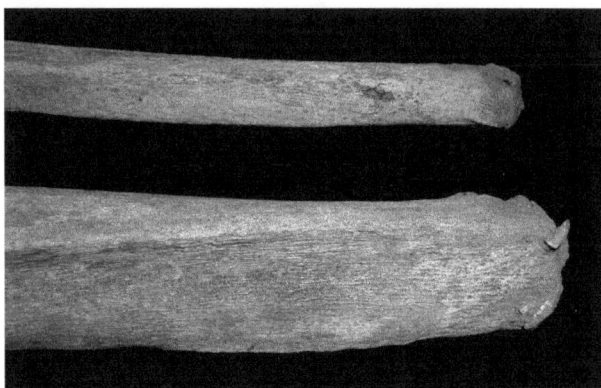

Figure 2. Close up of the left tibia and fibula from HB86 showing periosteal reaction.

The maximum tibial length was 241 mm compared to 382 mm for its right counterpart, showing a shortening of 141 mm. The maximum fibular lengths could not be compared due to the healthy one being broken post-mortem at its distal end. Although both severed bones terminated in a well-healed stub, they showed evidence of a long-term inflammatory process in the form of new bone formation affecting the diaphyseal surfaces of their shafts (Figure 2 & 3). The morphology of the cut surface, although healed, suggests that the operation was performed with a saw in a well-controlled surgical environment. Furthermore, the distal end of the tibia had a small osteophyte formed antero-laterally which, according to Barber (1929; 1934), tends to appear near the site of the mutilation as a consequence of the initial tearing of the periosteum. There was no evidence of disuse atrophy so a prosthesis may have been employed, although this was not found in the grave.

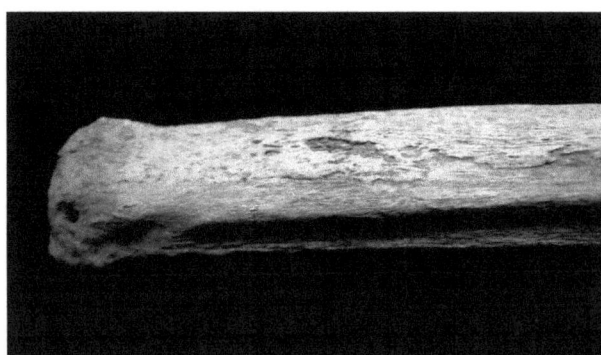

Figure 3. Left fibula from HB86 showing a "cloaca" suggestive of infection.

HB129 was a mature adult male (46+). It was not possible to calculate his stature as there were no complete long bones present. This individual had an amputation of the right femur (Figure 4) demonstrating a full recovery following surgical intervention with a well-healed stump. Judging by the difference in diameter between the femoral shafts, the loss of the limb had occurred a long time before death.

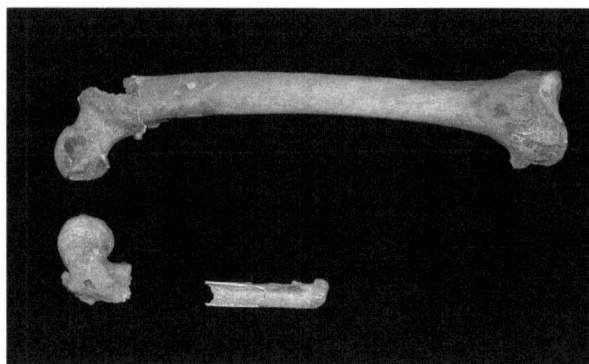

Figure 4. Both femora from HB129 illustrating the amputated fragment of the right femur.

Severe atrophy due to disuse of the amputated leg is manifested in the reduction in size of muscle attachment insertions on the *linea aspera*, in addition to the general remodelling of the entire diaphysis. Measurements taken antero-posteriorly at the nutrient foramen also corroborate this loss of cortical bone. The diameter of the amputated femur was 19.6 mm, much thinner than its normal counterpart, which was 28.3 mm. However, it is difficult to assess where exactly transection of the femur took place because only two broken pieces of this bone were found - a portion of diaphysis and the head and neck, leaving little indication of the original length of the amputated femur. There was also a well-healed fracture in the neck of the amputated femur (Figure 5), which may have occurred at the same time as the injury that necessitated amputation above the knee. The femoral neck fracture was in reasonably good alignment and there was no secondary joint disease in either the femoral head or acetabulum. The disuse atrophy of the stump suggests that a prosthesis was not employed since this would have kept the limb in a more active state.

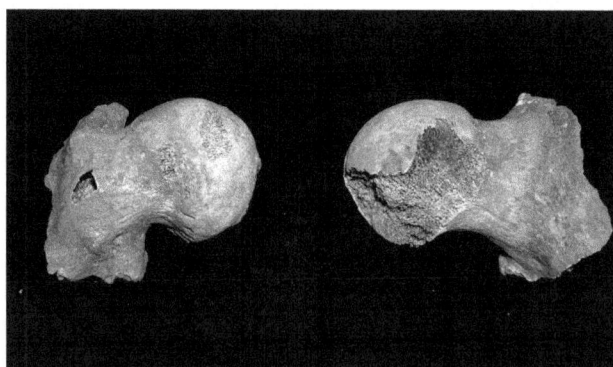

Figure 5. Left and right femoral head and neck of HB129. There is a healed fracture in the right femoral neck.

Finally, HB53 was a female individual and also a mature adult (46+) with an estimated stature of 1.58 m ± 3.72 cm. She appeared to have lost her left forearm just below the elbow a long time before death (Figures 6 and 7).

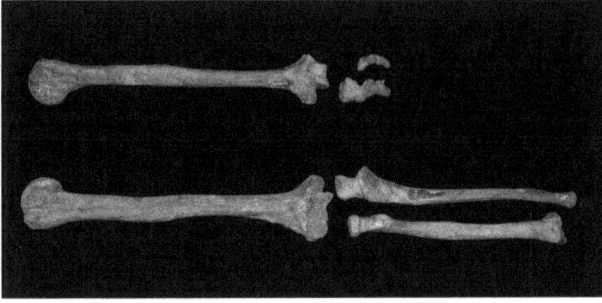

Figure 6. Upper limb from HB53.

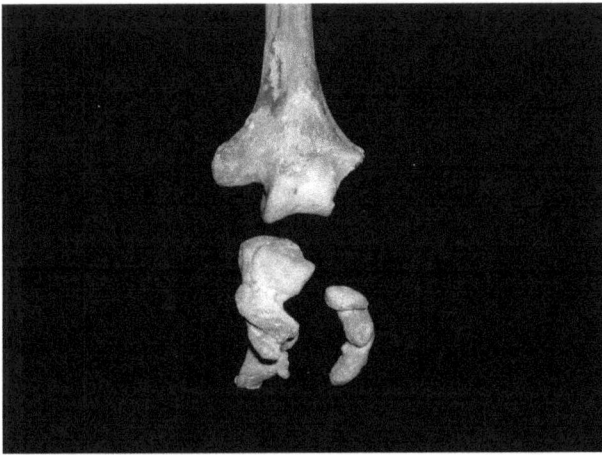

Figure 7. Close up of the amputation from HB53.

Small remnants of the left radius and ulna, the former measuring 30 mm in length and the latter 49 mm, were found. These showed a considerable amount of atrophic change and remodelling (Figures 6 and 7). No signs of infection were found in the stump or the rest of the arm. The lack of a cut surface and the shape of the atrophic bones suggest that the forearm was torn off as the result of a traumatic incident.

A comparison with the intact right radius (220 mm) and right ulna (241 mm) suggested that nearly 20 cm of both bones were missing. Drastic asymmetries in shape and size were found between the humeri, clavicles and scapulae reflecting how active the healthy arm was throughout life. In HB53, removal of the left hand and all its associated musculature would have led to a reduced strain and consequent diminution in size of the elbow and upper arm bones. Compensatory hypertrophy was seen in the deltoid insertion on the right humerus.

Discussion

The three cases of amputation from St. Peter's Church overflow cemetery are of great interest as they showed evidence of healing and survival after the operation, and provide direct evidence for this medical procedure having been performed in Wolverhampton during the first half of the 19th century. Although none of the individuals concerned was associated with a coffin plate, the cemetery was only in use from 1819 to 1850. It is notable that the local hospital, the South Staffordshire General

(later the Royal Hospital), was founded in AD1849 and prior to that the only medical facility in Wolverhampton was the Queen Street dispensary (Stallard 2005). It is recorded that an 18-year old girl had her foot amputated there on 1st January 1847 because it had been run over by a wagon. Dr E Hayling Coleman, the surgeon, was only the fourth person in Britain to use ether as the anaesthetic for this operation. The public dispensary at Queen Street had opened in 1821 as a six-bedded charitable institution and its proximity to St Peter's Church suggests that those buried in the overflow cemetery may have had their operations here.

The two male individuals analysed (HB86 and HB129) appeared to have been subjected to deliberate surgical amputation, although it is difficult to tell which of the two methods outlined above was used. Both cases demonstrated a clean cut, which would indicate that a qualified practitioner had performed the surgical intervention cautiously, being remarkably careful and precise as well as clean and fast. The presence of a small osteophyte at the distal end of the tibia in HB86 could be the direct result of the mutilation and the initial tearing of the periosteum as demonstrated by previous research (Barber 1929; 1934). Both cases had healed well with a rounded cap of bone having formed at the end of the stump and there was only slight evidence of residual inflammation in HB86.

In the case of HB129, it is difficult to tell whether the femoral neck fracture followed the amputation or occurred contemporaneously. According to Kostuik (1981) there are two events that may explain a fracture in an amputated limb. The first is that the amputation was due to an injury that also caused a fracture. The second is that an already-healed amputated limb suffered a subsequent fracture as a result of a new injury. In the case of HB129, the first explanation is the most likely one, since amputations performed on the leg are sometimes related to fractures of the femoral neck.

On the other hand, the only female individual of the group in question (HB53) appeared to have had her left forearm torn off, probably as a consequence of an accident (Manchester pers. comm.). It did not appear to have been surgically amputated since there was no intentional cut. As a result, the elbow joint was retained with two vestigial remnants of ulna and radius, which formed a small stump below the joint. As some of the parishioners of St Peter's Church are recorded to have worked in the local metal industries, it is possible that this woman caught her arm in a machine at her place of work. This case, therefore, is probably the result of an industrial accident.

Evidence of other amputations in the post-medieval period is provided firstly by the case of an adult male with an amputated left leg from Masshouse Circus, an urban cemetery located in Birmingham (Brickley et al. in press). He had died shortly after the amputation, probably due to an infection introduced during the surgical procedure. Secondly, a male with leprosy from St

Marylebone churchyard in London had had his right foot amputated, possibly as a consequence of the disease (Miles et al. in prep). Additionally, at the Royal Hospital in Greenwich, London five cases of amputation were identified in 104 individuals. One of these was conducted on the hand, another on the tibia and the remaining three on the femur (Witkin pers. comm.). As this was a naval hospital devoted to the long-term care of seamen whose military occupation had been extremely hazardous (Galer 2002), it was to be expected that the number of amputations would be comparatively higher than in the Wolverhampton sample.

By contrast, none of 857 individuals recovered from St. Martin's churchyard in Birmingham and dating to the 18th and 19th centuries was an amputee (Brickley et al. 2006). In addition, there were no cases of amputation among the 227 people buried in St. Bride's church crypt, Fleet Street, London that date to the same period (Scheuer 1998). A total of 968 individuals were buried in the crypt of Christchurch, Spitalfields, London, and although there was evidence of several types of fracture, no case of amputation was found (Waldron 1993). The occupations and lifestyle of the people buried in these locations were clearly less dangerous than those of the group from St Peter's.

The pathological changes which might have led to some amputations in the post-medieval period can be discerned from surgical waste recovered from the Newcastle Infirmary. This was a 'voluntary hospital' dating to the 18th and 19th centuries. Excavation of the burial ground in 1997 produced more than 200 amputated limbs among the disarticulated remains of 400 individuals, in addition to 210 articulated burials (Witkin 1997). Infections, both specific and non-specific, were the main causes of amputation in this sample and included a syphilitic tibia and a tuberculous knee joint. Other bones showed evidence of post-mortem intervention allowing surgeons to practise the amputation technique before trying it on a patient (Chamberlain 1999).

Conclusion

The population of Wolverhampton rose from approximately 18,000 in 1821 to 50,000 in 1845 as a result of industrialisation during the period in question (Stallard 2005). St Peter's Church is situated in the centre of the city and served a cross-section of the demographic. As well as the professional classes, evidence from coffin plates shows that those buried in the overflow cemetery practised various trades, such as grocery, or were employed in the local industries. The most relevant industry in Wolverhampton was metalworking. According to Walvin (1982) the most unpleasant and hazardous working conditions during the Victorian period were found in the metal, earthenware, glass and pottery factories. It is possible to speculate, therefore, that one or more of the amputees from St Peter's Church sustained their injury at work.

The founding of the Queen Street Dispensary in 1821 and the South Staffordshire General Hospital in 1849 meant that surgical treatment became available to a larger cross-section of the population owing to the charitable nature of these foundations (Stallard 2005). Previously such care had only been available to the privileged few since access to medical treatment was elitist and only affordable for wealthy people (Wood & Woodward 1984).

To conclude, it is certain that the injuries, at least in HB129 and HB53, took place a long time before death. According to Lazenby and Pfeiffer (1993), a fully healed stump and evidence of atrophy in the affected limb would support this. The atrophy and asymmetry due to disuse are remarkable in both cases as several muscle groups from the amputated limbs appear to have withered. This in turn has led to a uniform reduction in the amputated limb and is manifested in the overall size and cross-section, which are all smaller than normal due to inhibited periosteal remodelling.

Acknowledgements

The authors are grateful to Jo Adams and Birmingham University Archaeology Unit for the opportunity to study the very interesting collection from St Peter's Church, Wolverhampton and for permission to report the three amputations in the present study. We also thank Mr Roy Stallard TD for providing the very helpful information on the history of Wolverhampton hospitals and St Peter's Church. Our grateful thanks are also due to the two referees for their constructive criticisms and suggestions for the paper.

Literature Cited

Adams J and Colls K (2006) "Out of darkness, cometh light": life and death in 19th century Wolverhampton. Birmingham Archaeology Monographs Series 2. British Archaeological Reports, British Series. BAR Publishing: Oxford.

Aldea PA and Shaw WW (1986) The evolution of the surgical management of severe lower extremity trauma. Clinics in Plastic Surgery 13: 549-569.

Barber CG (1929) Immediate and eventual features of healing in amputated bones. Annals of Surgery 90: 985-992.

Barber CG (1934) Ultimate anatomical modifications in amputation stumps. Journal of Bone and Joint Surgery 16: 394-400.

Bass WM (1987) Human Osteology: a Laboratory and Field Manual. Missouri Archaeological Society: Missouri.

Bell C (1821) Illustrations of the great operations of surgery, trepan, hernia, amputation, aneurism and lithotomy. London.

Brickley M, Berry H and Western G (2006) The people: physical anthropology. In M Brickley and S Buteux (eds.): St. Martin's uncovered, investigations in the churchyard of St. Martin's-in-the Bull Ring, Birmingham. Oxbow Books: Oxford: 119-130.

Brickley M, Rudge A and Krakowicz R (in press) Birmingham's earliest recorded amputee? Archaeological observations at Masshouse Circus, Birmingham city centre. Birmingham and Warwickshire Transactions.

Brooks S and Suchey JM (1990) Skeletal age determination based on the os pubis: a comparison of the Acsadi-Nemeskeri and Suchey-Brooks methods. Hum an Evolution 5: 227-238.

Buisktra JE and Ubelaker DH, eds. (1994) Standards for Data Collection from Human Skeletal Remains. Proceedings of a Seminar at the Field Museum of Natural History. Indianapolis: Arkansas Archaeological Survey Research Series 44.

Chamberlain A (1999) Teaching surgery and breaking the law. Evidence for early dissection. British Archaeology, October: 48.

Duckworth T (1980) Lecture notes on orthopaedics and fractures. Oxford: Blackwell scientific publications.

Galer DA (2002) A biocultural and comparative analysis of fracture-trauma at Greenwich Naval Hospital: 1749-1857. University of Bradford. Unpublished MSc Dissertation.

Haeger K (1988) The illustrated history of surgery. Harold Starke: London.

Isçan MY, Loth SR and Wright RK (1984) Age estimation from the rib by phase analysis: white males. Journal of Forensic Sciences 29: 1094-1104.

Isçan MY, Loth SR and Wright RK (1985) Age estimation from the rib by phase analysis: white females. Journal of Forensic Sciences 30: 853- 863.

Kostuik JP (1981) Amputation, surgery and rehabilitation: The Toronto experience. Churchill Livingstone: New York.

Krogman WM and Iscan MY (1986) The human skeleton in forensic medicine, 2nd edition. Charles C Thomas: Springfield.

Lazenby RA and Pfeiffer SK (1993) Effects of a nineteenth century below-knee amputation and prosthesis on femoral morphology. International Journal of Osteoarchaeology 3: 19-28.

Lovejoy CO, Meindl RS, Przybeck TR and Mensforth (1985) Chronological metamorphosis of the auricular surface of the ilium. A new method for the determination of adult skeletal age at death. American Journal of Physical Anthropology 68: 1-14.

Mays SA (1996) Healed limb amputations in human osteoarchaeology and their causes: a case study from Ipswich, UK. International Journal of Osteoarchaeology 6: 101-113.

Meindl RS and Lovejoy CO (1985) Ectocranial suture closure: a revised method for the determination of skeletal age at death based on the lateral-anterior sutures. American Journal of Physical Anthropology 68: 57-66.

Merbs C (1989) Trauma, in M-Y Isçan and KAR Kennedy (eds): Reconstruction of life from the skeleton. Alan R Liss: New York: 161-89.

Miles A, Powers N and Wroe-Brown R (in prep) St Marylebone Church and burial ground: excavations at St Marylebone Church of England School, 2005. MoLAS Studies Series.

Neilson C and Coates G (2002) Excavations in advance of the extension to the Harrison Learning Centre, University of Wolverhampton, West Midlands. 2001. Post-excavation assessment and updated project design. Birmingham University Field Archaeology Unit. Project No 846.

Porter R (1997) The greatest benefit to mankind: a medical history of humanity from antiquity to the present. London: Fontana Press.

Roberts C and Manchester K (2005) The archaeology of disease. Alan Sutton: London

Roberts C and Cox M (2003) Health and disease in Britain: from prehistory to the present day. Sutton Publishing: Stroud.

Scheuer L (1998) Age at death and cause of death of the people buried in St. Bride's Church, Fleet Street, London. In M Cox (ed): Grave concerns: death and burial in England 1700-1850. CBA Research Report 113, Council for British Archaeology: York: 100-111.

Stallard R (2005) Wolverhampton hospitals' heritage. Wombourne Printers: Wolverhampton.

Stewart TD (1979) Essentials of forensic anthropology: especially as developed in the United States. Charles C Thomas: Springfield (IL).

Trotter M (1970) Estimation of stature from intact long bones. In TD Stewart (ed.): Personal Identification in Mass Disasters. Washington: Smithsonian Institution. pp71-83.

Ubelaker DH (1989) Human skeletal remains: excavation, analysis, interpretation. Taraxacum Press: Washington.

Waldron HA (1993) The Health of the Adults. In The Spitafields Project, Volume 2: the anthropology edited by Molleson T and Cox M. The Middling Sort. CBA research Report 86, Council for British Archaeology pp 67-87.

Walvin, J (1982) A child's world: a social history of English childhood, 1800 – 1914. Penguin: Harmondsworth.

Wangensteen OW and Wangensteen SD (1967) Some highlights in the history of amputation reflecting lessons in wound healing. Bulletin of the history of medicine 41: 97-131.

Wangensteen OW and Wangensteen SD (1978) The rise of surgery. W. Dawson & Sons Ltd: Folkestone.

Witkin A (1997) The cutting edge: aspects of amputations in late 18th and early 19th century. Sheffields: University of Sheffield, Unpublished MSc Thesis.

Wood R and Woodward J (eds) (1984) Urban Disease and Mortality in Nineteenth-Century England. Batsford: London.

Unusual Neck Pathology in a Nevisian prehistoric individual

Prehistoric Nevisian neck pathology

Sonia R Zakrzewski and Elaine L Morris

School of Humanities (Archaeology)
Avenue Campus
University of Southampton
Southampton SO17 1BF
UK
Tel: 023 8059 4778
Fax: 023 8059 3032
e-mail for correspondence: srz@soton.ac.uk

Abstract

Over the years, a small series of skeletal remains have been collected by the Nelson Museum. Despite the poor preservation of the material, several specimens exhibited pathological changes. Severe neck pathology was found in one specimen deriving from the prehistoric site of Coconut Walk. This paper presents the changes in the vertebrae and elsewhere within this individual and proposes block vertebrae (possibly even Klippel-Feil syndrome) as the diagnosis.

Keywords: Prehistoric, Nevisian, neck pathology

Introduction

Over many years, a series of skeletal remains have been recovered from the Caribbean island of Nevis (a small member of the Leeward Islands of the Caribbean). Most of this material is currently curated in the Nelson Museum on the island and consists of skeletal material collected from around the island by interested parties. Apart from one individual, none of the material has been obtained by direct excavation, and thus most are considered as isolated finds within uncertain archaeological associations. The remains of at least twenty-two individuals were studied in 2003. Due to the poor preservation, most of the individuals could not be assigned to a sex and could only be assessed to be "adult". Nevertheless several individuals showed evidence of pathological change, with one being noted for its unusual neck pathology, as described here. This paper presents the changes in the vertebrae and elsewhere upon this individual and proposes the diagnosis of block vertebrae (and potentially Klippel-Feil syndrome).

Nevis in Context

The small island of Nevis is only 90 square kilometres in size and lies near the top of the Lesser Antilles archipelago, about 300 kilometres south-east of Puerto Rico, and immediately west of Antigua (see Fig. 1). The island is littered not only with ruins from the sugar plantation period, but also with both ceramic and aceramic prehistoric sites. The colonial period intensive cultivation of sugar radically transformed Nevisian vegetation, thereby rendering it difficult to reconstruct the prehistoric ecosystem and lifeway (Wilson, 1989: 429). The island appears to have been initially colonised by hunter-gatherers at about 1000 BC, with several later waves of migration by horticulturalists. Ceramic prehistoric culture in the Lesser Antilles appears in the last few centuries BC (Saladoid), with later prehistoric culture (Ostionoid) appearing by about AD 600 (Wilson, 1989: 430-1). This Ostionoid culture is marked by changes in the zooarchaeological material recovered, which is hypothesised to be the result of either overexploitation of the local environment or of technological innovation (Wing & Scudder 1983) and the standardisation of the ceramic material (Wilson 1989).

Material and Methods

The skeleton discussed here was recovered from the Ostionoid period prehistoric settlement at Coconut Walk (Nevis site code JA-1), radiocarbon dated elsewhere on Nevis to AD 745 ± 135 (Wilson, 1989: 436).

The skeleton, CNW B&E, was collected by John Enwema, and given to the Nevis Historical and Conservation Society, and the material is currently stored at the Nelson Museum, in a box labelled "Skeletal material from Coconut Walk". The skeletal material was studied by the first author (SRZ). The material was stored as two separate individuals (B and E) but was reunited after inspection. Individual B consisted only of a fragmented cranium and cervical vertebrae, whereas individual E consisted only of postcrania (including thoracic & lumbar vertebrae).

For all material studied, including the specimen CNW B&E, sexing was undertaken using traditional pelvis and skull characteristics (following Brickley and McKinley (2004) and Buikstra & Ubelaker (1994)). Individuals were classified on the 1 to 5 scale from definite male, through potential male, unsexed, and potential female, through to definite female. As all material studied was of adults, age was determined by tooth wear (Brothwell 1981), pubic symphysis morphology (Brooks & Suchey

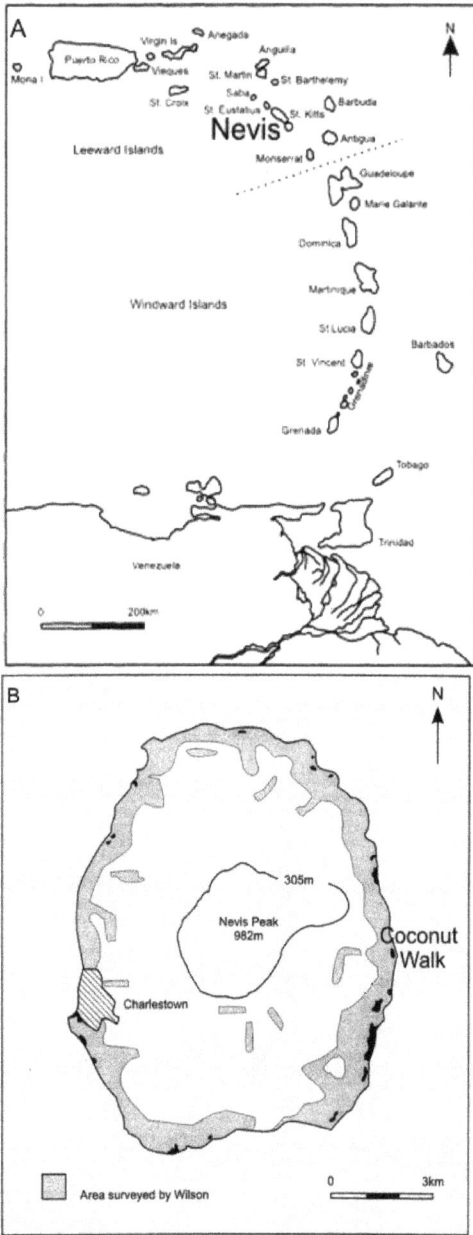

Figure 1. Location of Nevis and Coconut Walk (A: map of the Lesser Antilles; B: Map of Nevis with archaeological sites marked in black). [modified after Crosby (2003: 8-9)].

1990), auricular surface morphology (Lovejoy et al. 1985) and cranial suture closure (Perizonius 1984). As the material was fragmentary, individuals were simply classified into broad age bands (young adult, middle adult and older adult).

Description of material

CNW B&E consisted of cranial fragments, vertebrae and some other postcrania. Excluding the cervical vertebrae, all were highly fragmented. All cervical vertebrae were complete. The atlas exhibited some elongation of the articular facets associated with lipping of the superior aspect (see Fig. 2). The axis was completely fused with C3. Both the bodies and neural arches were entirely

fused together. The dens was unaffected. Severe osteophytic lipping was found on the anterior margin of the inferior surface of the body of C3 (see Figs. 3 & 4). C4 was severely affected as the body is highly curved and thinned, with increased macroporosity and associated with osteophyte development on the anterior margin (Fig. 5). C5 exhibited similar pathological changes. Overall, the cervical vertebrae exhibited medio-lateral scoliotic displacement, with the lower vertebrae being displaced towards the individual's right.

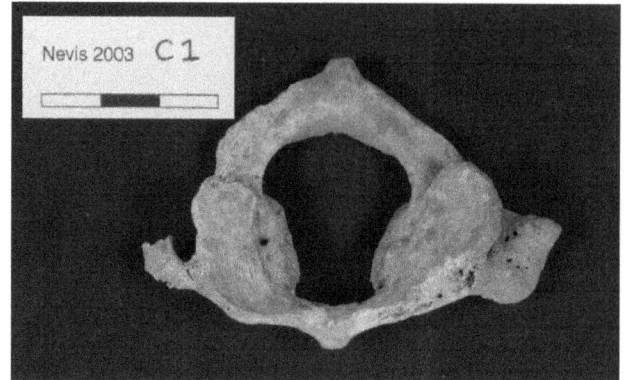

Figure 2. Superior view of C1.

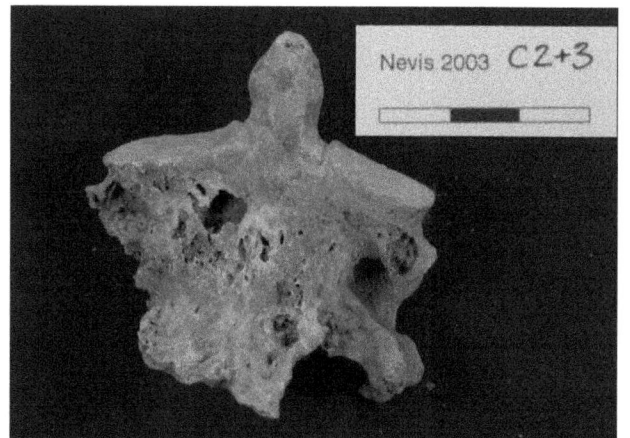

Figure 3. Anterior view of C2 & C3.

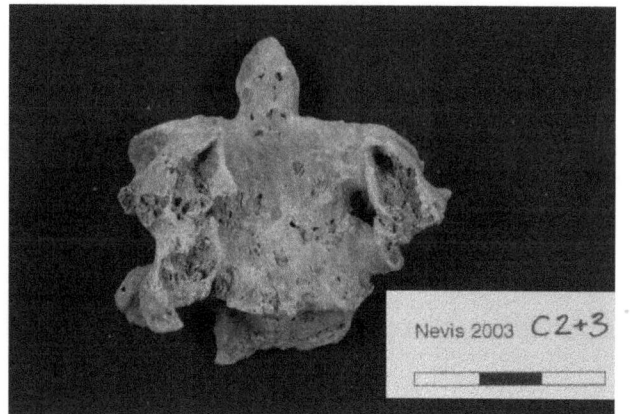

Figure 4. Posterior view of C2 & C3. Note osteophytic lipping of the inferior surface of C3.

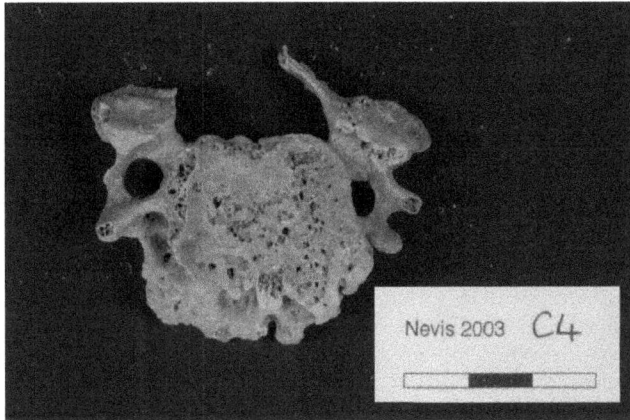

Figure 5. Superior view of C4. Note osteophytic lipping of the anterior-posterior margin

The remains of at least ten thoracic vertebrae are preserved, with the inferior-most two having osteophyte development. The remains of all five lumbar vertebrae were noted, with all expressing pathological changes. All have compression of the bodies and osteophytic lipping. Furthermore, an additional highly fragmentary lumbar vertebra was noted, likely due to lumbarisation of T12 having occurred. Given the highly fragmentary nature of this vertebra, it is possible that it is an additional lumbar or sacral vertebra, but lumbarisation is more likely (Barnes 1994: 78-9).

Eighteen rib fragments were associated with these vertebrae, of which seven could be sided (four left and three right). Lipping was found on many of the fragments, with osteophyte development inside the single costal end preserved. Unfortunately, very little pelvic material remains. The rest of the postcrania consisted of the body of the sternum, a fragmentary unsided glenoid fossa, a partial unsided humerus head and shaft, both right and left shafts and distal epiphyses of the ulnae, a fragmentary acetabulum, the medial portion of the distal articular surfaces of the left femur, the right patella and an unsided fibula midshaft. The teeth all exhibit low to moderate wear and the anterior central incisors exhibit and unusual wear pattern. This may be either accidental angled dental wear or deliberate shaping.

Age and Sex

Little suitable skeletal material for age determination was preserved. All teeth exhibited low to moderate wear, and hence this individual was provisionally classified as being a young adult through Brothwell's (1981) method.

Unfortunately, neither the cranium nor the os coxae was sufficiently complete to permit sex determination. Furthermore the postcranial material that was preserved was insufficient to allow metric sex determination methods. The maximum height of the right patella was 40mm, midway between white South African male and female values (Bidmos et al. 2005); hence this specimen remains unsexed.

Differential Diagnosis

The fusion and scoliosis in the cervical vertebrae may be caused by traumatic injury to the neck associated with healing, *spondylosis deformans*, Klippel-Feil syndrome or due to an error in segmentation leading to the development of block vertebrae. These will each be considered below.

Traumatic injury

Trauma was suggested by the complete fusion of the C2 and C3 vertebrae and the associated osteophyte development and macroporosity on the rest of the cervical vertebrae (Ortner & Putschar 1981). No complete fracture was found, therefore a potential traumatic injury would be infraction to either C2 or C3 with a prolonged period of bony growth. In order to form the scoliosis observed, the force required would need to be bending in nature and affect both vertebrae. It was not possible to X-ray the specimen and, from visual examination, no macroscopic fracture line was found. Furthermore no callus was noted. If fracturing had occurred, the episode must have occurred long before the death of the individual as the vertebrae were fully fused and healthy. Furthermore, for fusion to occur, the vertebrae would likely need to have become immobilised together. However, this diagnosis, potentially associated with spondyloarthropathy, is supported by the lower vertebral compression and osteophyte development. Traumatically-induced fusion of the vertebrae may occur as a result of compression injuries, such as those resulting from hyper-flexion of the neck (Resnick & Niwayama 1981). This potential cause of the palaeopathology is considered unlikely in the Nevisian population.

Spondylosis deformans

Spondylosis deformans may occur in any segment of the vertebral column. The disease is characterised by bridging osteophytes, arising from the junction of the body and the fused marginal vertebral plate which bulges at the level of the intervertebral disc. This is the most common degenerative spinal disease to affect males in greater proportion than females (Mann & Murphy 1990: 57). Over a prolonged period, remodelling may give the bridges a smooth appearance, thereby making the condition resemble ankylosing spondylitis[1] (Ortner & Putschar 1981: 421).

This diagnosis is supported by the proliferation of osteophytes elsewhere within the vertebral column, the macroporosity noted in the cervical vertebral bodies and the potentially age-related compression the lumbar bodies. It is unlikely that the complete fusion would occur in only two vertebrae, especially C2 and C3,

---

[1] Ankylosing spondylitis itself was not considered a possible diagnosis as the condition starts in the lower back and sacroiliac region, and then moves up the vertebral column, whereas the fusion noted affects only C2 and C3.

without other fusion of vertebrae occurring unless this is an early stage. The diagnosis of *spondylosis deformans* is unlikely as there is no inflammation observable, and hence it would be required to be secondary to trauma or related to a bone-forming degenerative condition such as diffuse idiopathic skeletal hyperostosis (DISH) or ankylosing spondylitis.

Block vertebrae

The block formation of two adjacent vertebrae occurs as a result of incomplete segmentation of the vertebrae. The fusion may be a result of congenital absence of the intervertebral disc (Aufderheide & Rodríguez-Martin 1998: 62-3) due to caudal shifting during the sclerotome phase of development (Barnes 1994). Clinical incidence rates are reported to be 2-4%, with the most commonly affected vertebrae being in the cervical spine (Aufderheide & Rodríguez-Martin 1998) and lumbar regions. Complete unity is found between the centra and is associated with either complete or incomplete unity of the neural arches, although, usually the neural arches are affected (Barnes 1994: 66-7).

For this individual, this diagnosis is supported, as the vertebrae affected are C2 and C3, with no separation between the bodies and with associated fusion of the transverse processes and pedicles.

Klippel-Feil syndrome

This condition, also known as brevicollis (Kaplan et al. 2005: 573), consists of the "congenital fusion"[2] of two or more vertebral segments into a block vertebra with a single spinous process, neural arch and vertebral body. This is the result of block vertebrae forming due to segmentation failure in embryogenesis (Aufderheide & Rodríguez-Martin 1998: 60). The neck thus becomes shortened and has limited mobility (Barnes 1994: 67). The prevalence is less than 0.01%, and so thus may be considered as a specialised form of block vertebrae.

Three types have been defined (Barnes 1994: 69):

Type I – involving several cervical and thoracic vertebrae forming one grossly abnormal osseous block,
Type II – involving only 2 or 3 vertebrae, with C2 and C3 being most common, and believed to be an autosomally recessive trait, and,
Type III – involving cervical vertebrae and associated with other segment errors in the thoracic and lumbar regions.

The type II form is the most common and is usually asymptomatic.

With this syndrome, the vertebral bodies become flattened or widened and the disc space is reduced. The

syndrome is also associated with scoliosis (especially of the cervical vertebrae), malformation of the occipital, elevation of the scapula, spina bifida, cleft palate, extra cervical vertebrae, fusion of the ribs, facial asymmetry and torticollis (Barnes 1994; Douglas 1991).

Klippel-Feil type II is a potential diagnosis of the pathology noted in the individual above. The specimen has C2 and C3 completely united, with associated scoliosis. Unfortunately, the occipital is very poorly preserved and so cannot be assessed. Only the portions of the left side of the maxilla are preserved and so cleft palate cannot be assessed. Only one fragmentary glenoid of the scapula and no sacral material were preserved, hence the other typically associated pathologies cannot be assessed.

Discussion

The evidence presented suggests that the upper cervical pathology described arose from type II Klippel-Feil syndrome as a result of block vertebral formation of C2 and C3. In living persons, individuals exhibiting Klippel-Feil syndrome are noted for having short necks, low posterior hairlines, limited neck mobility and occasional skin webs from the neck to the shoulder (Riseborough & Herndon 1975: 197). Although believed to be the result of an autosomally recessive trait (Barnes 1994: 69), potentially through the PAX1 gene locus on the long arm of chromosome 8 (McGaughran et al. 2003), Klippel-Feil syndrome is a heterogeneous disorder, exhibiting different expressions in different families (González-Reimers et al. 2001).

Archaeologically, Klippel-Feil syndrome has been noted among the Anasazi Pueblo groups (Barnes 1994: 69-71), individuals from prehistoric Peru (Ortner & Putschar 1981: 357), prehistoric Mexico, Ptolemaic or Late Period Egypt (Aufderheide & Rodríguez-Martin 1998: 60), prehispanic Canary Islanders (González-Reimers et al. 2001), and Austrian Magyar period groups (Pany et al. 2004).

The individual also appears to have had "lumbarisation" of T12. This is associated with cranial shifting during embryogenesis (Barnes 1994: 104-109), and may have a similar aetiology to the C2 and C3 block vertebrae.

The impact upon the individual's life is hard to predict. As torticollis and facial asymmetry affect almost half of sufferers, it is possible that this individual looked different to other people in the population. If this is the case in the Coconut Walk individual, this may have led to the differential treatment by the population. Unfortunately, this hypothesis cannot be tested as the cranial remains were too fragmentary to study, and the specimen itself derives from a potentially disturbed context. Neurological and renal problems occur in many individuals with the disorder (Sullivan 2005), and hearing loss is common in about one third or sufferers (Hensiger et al. 1974; Kaplan et al. 2005: 574). It is, therefore, possible that this individual appeared in some

---

[2] This is not truly fusion of the two vertebrae as the segments never actually separated. For detail regarding embryogenesis of the vertebrae, see Barnes (1994), Kaplan et al. (2005) or Scheuer & Black (2000).

way different to the rest of the local population and hence may have acted or been treated differently.

Conclusion

An individual has been described who has severe pathological changes to the vertebral column, especially the upper vertebral column. These changes consist of "fusion" of C2 and C3, cervical scoliosis, increased cervical macroporosity and osteophytosis. These are associated with osteophyte development in the lower thoracic and lumbar regions and lumbarisation of T12. Although it is possible that this pathology results from traumatic insult, we argue that the most likely diagnosis is type II Klippel-Feil syndrome. This is an autosomally recessive trait, and thus, if found in other prehistoric Nevisian (or other Caribbean) islanders, may be used to aid in genetic population reconstruction.

Acknowledgements

The work was undertaken as part of the Nevis Heritage Project of the University of Southampton, in association with the Nevis Historical and Conservation Society and the Bristol and Region Archaeology Service. The authors would like to thank the Nelson Museum for access to the material and to Nevis Historical and Conservation Society for their commitment to the curation of disturbed skeletal material found on the island, and to Penny Copeland for the photographs. The Nevis Heritage Project has been in grateful receipt of grants from both the British Academy and the Society of Antiquaries of London.

Literature Cited

Aufderheide AC and Rodríguez Martín C (1998) The Cambridge Encyclopedia of Human Paleopathology. Cambridge, UK: Cambridge University Press.

Barnes E (1994) Developmental Defects of the Axial Skeleton in Paleopathology. Niwot: University Press of Colorado.

Bidmos MA, Steinberg N, and Kuykendall KL (2005) Patella measurements of South African whites as sex assessors. Homo 56:69-74.

Brickley M and McKinley JI, eds. (2004) Guidelines to the Standards for Recording Human Remains. Reading: Institute of Field Archaeologists No. 7.

Brooks S and Suchey JM (1990) Skeletal age determination based on the os pubis: A comparison of the Acsadi-Nemeskeri and Suchey-Brooks methods. Human Evolution 5: 227-238.

Brothwell DR (1981) Digging Up Bones: The Excavation, Treatment and Study of Human Skeletal Remains. London: British Museum.

Buikstra JE and Ubelaker DH, eds. (1994) Standards for Data Collection from Human Skeletal Remains. Fayetteville, Arkansas: Arkansas Archeological Society.

Crosby A (2003) The Prehistoric Communities: The Hichman's Landscape. In E Morris, R Leech, A Crosby, T Machling, B Williams and J Heathcote (eds.): Nevis Heritage Project: Interim Report 2002. Southampton: University of Southampton.

Douglas MT (1991) Wryneck in the ancient Hawaiians. American Journal of Physical Anthropology 84:261-71.

González-Reimers E, Mas-Pascual A, Arnay-De-La-Rosa M, Velasco-Vasquez J and Jimenez-Gomez MC (2001) Klippel-Feil syndrome in the prehispanic population of El Hierro (Canary Islands). Annals of the Rheumatic Diseases 60:174.

Hensinger RN, Lang JE and MacEwen GD (1974) Klippel-Feil syndrome; a constellation of associated anomalies. The Journal of Bone & Joint Surgery (A) 56:1246-1253.

Kaplan KM, Spivak JM and Bendo JA (2005) Embryology of the spine and associated congenital abnormalities. The Spine Journal 5:564-576.

Lovejoy CO, Meindl RS, Mensforth RP and Barton TJ (1985) Multifactorial determination of skeletal age at death: a method and blind tests of its accuracy. American Journal of Physical Anthropology 68:1-14.

Mann RW and Murphy SP (1990) Regional Atlas of Bone Disease. Springfield: Charles C Thomas.

McGaughran JM Oates A, Donnai D, Read AP, and Tassabehji M (2003) Mutations in PAX1 may be associated with Klippel-Feil syndrome. European Journal of Human Genetics 11:468-474.

Ortner DJ and Putschar WGJ (1981) Identification of Pathological Conditions in Human Skeletal Remains. Washington: Smithsonian Institution Press.

Pany D, Teschler-Nicola M, Kainberger F, Prohaska T and Stingeder G (2004) Klippel-Feil Syndrome in a Magyar Conquest Period juvenile skeleton from Austria. Poster presented at the 15th European meeting of the Paleopathology Association, Durham.

Perizonius WRK (1984) Closing and Non-closing sutures in 256 Crania of Known Age and Sex from Amsterdam (A.D. 1883 - 1909). Journal of Human Evolution 13:201-216.

Resnick D and Niwayama G (1981) Diagnosis of bone and joint disorders. Philadelphia: WB Saunders.

Riseborough EJ and Herndon JH (1975) Scoliosis and other deformities of the axial skeleton. Boston: Little, Brown & Company.

Scheuer L and Black S (2000) Developmental Juvenile Osteology. London: Academic Press.

Sullivan JA (2005) Klippel-Feil syndrome. http://www.emedicine.com/orthoped/topic408.htm Accessed 15-26 Jan 2006.

Wilson SM (1989) The Prehistoric Settlement Pattern of Nevis, West Indies. Journal of Field Archaeology 16:427-450.

Wing ES and Scudder SJ (1983) Animal exploitation by prehistoric people living on a tropical marine edge. In C Grigson and J Clutton-Brock (eds.): Animals and Archaeology. Oxford: BAR International Series 183, pp. 197-210.

The Search for Rosa Pike: Congenital Syphilis in 1880s London

Congenital Syphilis in Victorian London

Roshni Patel & Piers D. Mitchell*

Faculty of Medicine
Imperial College London
84 Huntingdon Road
London N2 9DU
* e-mail address for correspondence: piers.mitchell@imperial.ac.uk

Abstract

In the nineteenth century the prevalence of venereal syphilis in London was such that newborn children ran a significant risk of developing congenital syphilis. The aim of this paper is to highlight this issue through investigation of the pathological and historical evidence for one such an infant, named Rosa.

The research focuses upon an osteological specimen, dating back to 1886, which is curated at the pathology museum of Imperial College London. The cranium is of a one year old child named Rosa Pike. Records described how her mother had extensive ulcerated nasofacial syphilis four years prior to Rosa's birth. The case notes recorded that Rosa developed a 'rash on her buttocks', which is common in congenital syphilis. Pathological analysis of her cranium demonstrates small destructive lesions on the frontal bones, craniotabes of the vault, a thick layer of periosteal bone on the parietal eminences, and hard, white sclerotic bone around the anterior fontanelle. All of these are established features of congenital syphilis.

In order to investigate Rosa's social circumstances, a study of records at the national Family Record Centre in Islington, London was undertaken. Her full name was Rosina Pike, this Christian name often being interchanged with Rosa in Victorian times. She lived with her working class mother and father in Fulham, then a poor area on the peripheries of London. This case exemplifies the risks to young children from congenital syphilis in Victorian London.

Keywords: congenital syphilis, treponemal disease, cranial lesions, historical records, malnutrition.

Introduction

The aim of this research is to highlight the consequences of congenital syphilis in Victorian London by discussing the case of one child with characteristic skeletal lesions of the disease. Palaeopathological and historical evidence has been used to investigate the infant, named Rosa Pike, who died in 1886.

In Victorian times venereal syphilis was well recognised as a health hazard to all social classes. The disease did not only have medical implications, but moral ones as well, as it was a venereal disease. Physicians of the day realised that the poor were at greater risk than the rich, because they could not afford medical treatment and many women were obliged to work as prostitutes in order to buy food (Hyde 1897). There were thousands of prostitutes working in London at any one time in the late nineteenth century. Some were full time, while others supplemented their family income with occasional work (Fisher 1997). A number of laws such as the Contagious Diseases Act were introduced, repealed, modified and reintroduced in attempts to manage the problem (Bartley 2000). Owing to the stigma associated with syphilis, poor people were often turned away from London's charitable hospitals if they had this disease, while others with non-venereal diseases were taken in for treatment (Acton 1851). One consequence of so much active venereal syphilis was that children were born with congenital syphilis. It is unknown exactly how many such children there were as Victorian records often avoided the diagnosis, on account of the associated stigma. The prevalence of congenital syphilis largely depends upon the stage of the disease in the mother. Those conceived when the mother had recently contracted the disease run a 50% risk of becoming infected themselves. Those conceived when the mother was in a latent period run just a 10% risk of contracting the disease. Roughly 25-50% of foetuses that do contract syphilis are lost from miscarriage or still birth (Adler 2002). However, this would still leave a large number of babies born alive with congenital syphilis in a city the size of Victorian London.

Method

The skull of Rosa Pike is curated in the Pathology Museum of Imperial College London. It is accompanied by the entry card that was completed after her post mortem. Since the record shows Rosa died in 1886, we presume the entry was completed at that time. The card certainly predates 1900 as the style of card changed at that time. The post mortem was performed at the Westminster Hospital, which is where the Pike family attended for Rosa's treatment while she was alive. The card not only records Rosa's symptoms, but also those of her mother. We recorded the lesions visible today on the skull, and compared these with the skull of a child of similar age who died from a different condition that left no skeletal lesions. We also compared the lesions with the clinical symptoms and signs from the Westminster Hospital appointments to determine if just one disease may have been responsible, or whether Rosa may have suffered with a number of diseases.

We investigated Rosa's social background by identifying her birth certificate and death certificate. This involved research at the Family Record Centre in Islington, London, which contains the national register of births, marriages and deaths in England.

Results

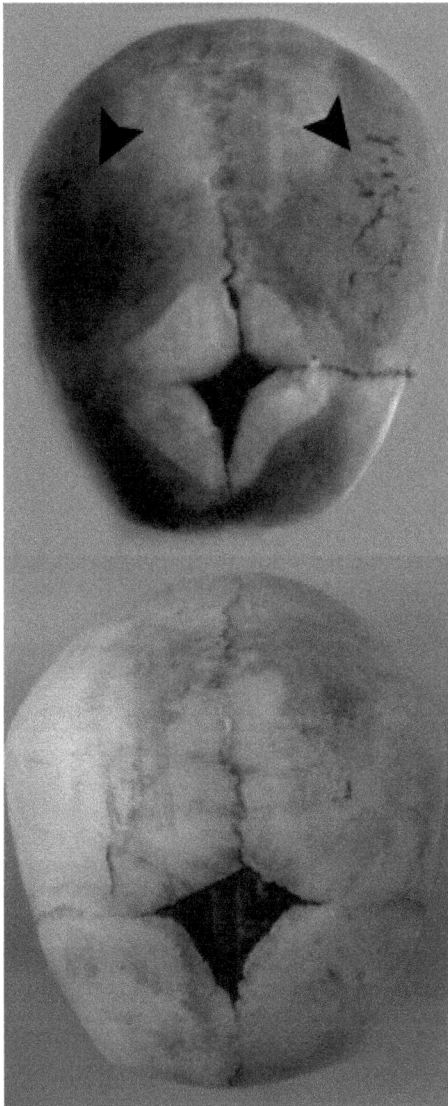

Figure 1a. (top). Superior view of the cranial vault of Rosa's skull. A porous layer of bone with vascular grooves present on the parietal and frontal bones.
Figure 1b. (bottom). Superior view of a normal infant, for comparison.

The specimen was noted to possess a number of abnormalities on the cranium that were not present on a normal skull of similar age. The shape of the skull is not normal, with bossing at parietal and frontal eminences. A

thickened layer of porous bone has formed over the parietal eminences (Fig.1a). This is grey in colour compared with the rest of the skull, which is pale cream in colour. Multiple grooves and pits are present in this thickened bone, suggestive of vascular markings. These are not present on the normal comparative skull (Fig. 1b). Away from these thickened areas the cranial vault thickness is abnormally thin (craniotabes). The contours of the gyri of the brain are clearly visible on the inner table of the skull. In parts of the parietal and occipital bones, the cortex is so thin that light easily shines thorough it, and a number of round perforations are present right through the cortex (Fig.2). These appear ante mortem in nature, and are associated with the thinnest sections of cortex. The anterior fontanelle remains wide open. It is bordered by a hard margin of bone that appears more dense than most of the skull, with the appearance of white marble. On both frontal bones there are localised areas of destruction of the outer cortex that do not involve the inner cortex (Fig.3). No new bone has formed at these lesions. These are irregular in shape, and appear to have a different aetiology to the smooth, round perforations noted in the thinned bone more posteriorly.

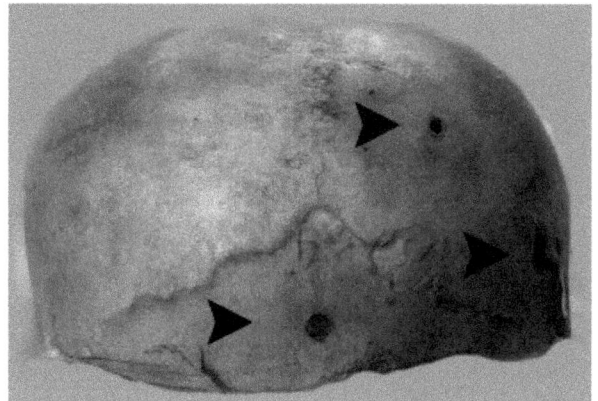

Figure 2. Posterior view of the cranium, antemortem round perforations

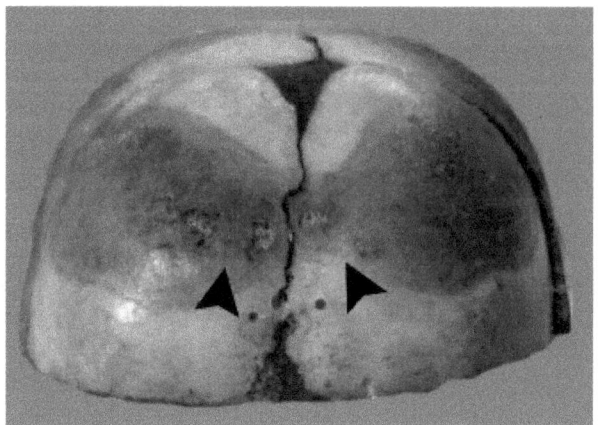

Figure 3. Frontal view of cranium, showing localised areas of lytic destruction on the outer table

The museum card records the symptoms and signs present in both Rosa and her mother (Fig.4). It described how Rosa's mother suffered with extensive ulceration of her nose and face four years before. This is highly suggestive of treponemal disease, and in the context of a 19th century British city the most likely form would have been venereal syphilis. The record also states that when Rosa was born she appeared healthy. In her second month of life she developed a skin rash on her buttocks. Breast feeding was performed for just one month, before weaning to a diet of condensed milk and Ridge's Food, a popular milk substitute for babies in Victorian Britain (Redmond & Hunter 1876). The doctors also noted that Rosa clinically had bossing of the skull when they saw her in March 1886 (aged 9 months). The record concludes by stating that the final cause of death was believed to be a chest infection.

Investigations at the Family Record Centre in Islington, London showed that in 1886, there was only one child aged one year registered with the name of Rosa Pike who died in the whole of England. She lived in Fulham, not far from the Westminster Hospital. Her birth and death certificates showed that she was born in May 1885 and died on 20 May 1886 (Fig.5). Her father was an omnibus washer named Henry. This confirms that the Pikes were a working class family. The death certificate actually records her name as Rosina while the medical record calls her Rosa. Study of Victorian books on Christian names shows that Rosina was the diminutive version of the name Rosa (Anonymous 1884; Swan 1900). Rosa was a name common in Italy and Spain, equivalent to Rose in England (Withycombe 1977). In Victorian times it became fashionable to use the variants Rosa and Rosina, so this explains why these two versions of the name appear in different documents.

History: Rosa Pike, agé. 1 year. The history of syphilis was complete. The mother had had extensive ulceration of the face and nose four years ago. At the time of birth the child was quite well, but a rash appeared on the buttocks at the second month. The child was suckled for one month and then fed on Ridge's Food and condensed milk. When seen for the first time, March 13, 1886, the bossing was well marked. The child died of pleuro-pneumonia.

Figure 4. Pathology museum entry card, recording the symptoms and signs of both Rosa and her mother in the 1880's.

CERTIFIED COPY OF AN ENTRY OF DEATH

GIVEN AT THE GENERAL REGISTER OFFICE

Application Number COL704163

REGISTRATION DISTRICT Fulham

1886 DEATH in the Sub-district of Fulham in the County of Middlesex

Columns:– 1		2	3	4	5	6	7	8	9
No.	When and where died	Name and surname	Sex	Age	Occupation	Cause of death	Signature, description and residence of informant	When registered	Signature of registrar
270	Twentieth May 1886 43 Waltham Avenue	Rosina Pike	Female	12 months	Daughter of Henry James Pike an Omnibus Washer	Marasmus Tetanus Certified by J C Fox M.B.	E Pike Mother Present at the Death 43 Waltham Avenue Fulham	Twenty first May 1886	David Shopland Registrar

CERTIFIED to be a true copy of an entry in the certified copy of a Register of Deaths in the District above mentioned.

Given at the GENERAL REGISTER OFFICE, under the Seal of the said Office, the 28th day of April 2005

DYA 660819

See note overleaf

CAUTION: THERE ARE OFFENCES RELATING TO FALSIFYING OR ALTERING A CERTIFICATE AND USING OR POSSESSING A FALSE CERTIFICATE ©CROWN COPYRIGHT

WARNING: A CERTIFICATE IS NOT EVIDENCE OF IDENTITY.

GENERAL REGISTER OFFICE ENGLAND

GJB

031998 8283 10/04 SPSL 009197

Figure 5. Copy of death certificate for Rosa Pike, dated 21st May 1886.

Discussion

A number of fascinating pieces of interlinking evidence have been presented here, namely a child's pathological cranium, clinical records from 1886, and documents identifying the social background of the individual. However, a number of points must be considered in order to safely draw conclusions from this information.

The clinical records do sound as though Rosa's mother had symptoms and signs likely to represent venereal syphilis. We can only speculate as to how she may have contracted this. She may have been a prostitute before marrying and having Rosa, she may have worked as a prostitute while married to supplement the family income, or she may have caught the disease from her husband. As not all children develop congenital syphilis even if their mother is infected, we must look for written evidence that Rosa may have the disease. The rash noted on Rosa aged one or two months old is common in congenital syphilis, and the location around the buttocks is also classical (Adler 2002). This was also recognised as a sign of the disease by physicians of the period Rosa was alive (Ballard 1874). The clinical observations support the conclusion that she did have the disease.

As for the osteological lesions noted on Rosa's skull, we must consider whether all were caused by congenital syphilis, or if she also had other diseases that may have left lesions. Scurvy, rickets and anaemia are all known to leave changes on the skull that may be seen in infants. Scurvy has been noted to cause porosity and hypertrophy in the greater wing of the sphenoid, with porosity in the orbital roof (Brickley & Ives 2006; Ortner & Ericksen 1997). However, the distribution and nature of the lesions in Rosa's skull are very different to this. Rickets can lead to cranial vault porosity and orbital roof porosity (Mays et al. 2006; Ortner & Mays 1998), and a combination of rickets with scurvy (Barlow's Disease) may cause thickening to the parietal bones similar to that seen in this specimen. However, only congenital syphilis gives the combination of destructive lesions of the outer cortex of the frontal bones, parietal bossing due to thickened hypervascular new bone, and thinning of bone elsewhere with craniotabes and perforations (Dennie & Pakula 1940). A limited number of archaeological cases of congenital syphilis have been published in the literature (Erdal 2006; Ferencz & József 1992; Gladykowska-Rzeczycka & Krenz 1995), so most of our evidence for its lesions comes from pathology museum collections. Indeed, there was much interest amongst the medical profession in the pathology of congenital syphilis around the time of Rosa's death in the 1880s (Fournier 1886; Kassowitz 1882; Taylor 1875), which may explain why she underwent post mortem examination in the first place.

It is also important that we consider the cause of death for this child. The death certificate recorded that she died from marasmus and tetanus. Marasmus is a form of malnutrition where insufficient energy is supplied by the diet (Hendricks and Duggan 2005: 49). Tetanus is an infection by the bacterium *Clostridium tetani* that produces a neurotoxin, leading to muscular spasms and death (Bleck 2000). There is no mention of congenital syphilis on the death certificate. It could be that she really did die from these conditions and that the syphilis was undiagnosed before the post mortem examination. Another possibility is that the condition was recognised, but for reasons of stigma associated with the disease the doctor wrote marasmus and tetanus on the form so as not to embarrass the family. A third option is that she did not have marasmus and tetanus at all, but that the effects of syphilis were mistaken by the certifying doctor as marasmus and tetanus. Congenital syphilis commonly causes growth impairment and failure to thrive that can mimic protein-energy malnutrition (Adler 2002). Furthermore, syphilitic meningitis may cause many of the symptoms and signs of tetanus. The cause of death on the Westminster Hospital records was a chest infection (pleuripneumonia). However, a recognised effect of congenital syphilis is pnemonitis (Adler 2002), where inflammation of the lungs would give very similar symptoms and signs to a childhood pneumonia. In consequence, it seems likely that the causes of death noted in the death certificate and hospital records were all manifestations of the congenital syphilis suffered by Rosa, and not separate conditions after all.

Conclusion

We have combined 19th century osteological material with medical records and social history documents in order to best recreate the short life of this unfortunate child. She only lived for one year, and in that time suffered with a range of presentations of congenital syphilis, including skin rashes, weight loss and lung disease. In this way we have highlighted the problem of congenital syphilis in Victorian London, and helped integrate the role of palaeopathology with the historical evidence to better understand one of the major childhood diseases of the 19th century.

Literature Cited

Acton W (1851) Prostitution in Relation to Public Health. London: John Churchill.

Adler SP (2002) Intrauterine infections; Treponema pallidum. In HB Jenson and RS Baltimore (ed.): Paediatric Infectious Diseases: Principles and Practice. Philadelphia: WB Saunders: 1102-7.

Anonymous (1884) The Pocket Dictionary of 1,000 Christian Names (Masculine and Feminine) with their Meanings Explained in Four Different Ways for Ready Reference, 2nd Edition. London: John Hogg.

Ballard T (1874) An Enquiry into the Value of the Signs and Symptoms Regarded as Diagnostic of Congenital Syphilis in the Infant. London: Churchill.

Bartley P (2000) Prostitution: Prevention and Reform in England, 1860-1914. London: Routledge.

Bleck TP (2000) Clostridium tetani (Tetanus). In GL Mandell, JE Bennett and R Dolin (eds.): Principles and Practice of Infectious Diseases. Philadelphia: Churchill Livingstone; 2537-43.

Brickley M and Ives R (2006) Skeletal manifestations of infantile scurvy. American Journal of Physical Anthropology 129: 163-72.

Dennie CC and Pakula SF (1940) Congenital Syphilis. Philadelphia: Lea and Febiger.

Erdal YS (2006) A pre-Columbian case of congenital syphilis from Anatolia (Nicaea, 13th century AD). International Journal of Osteoarchaeology 16: 16-33.

Ferencz M and Józsa L (1992) Congenital syphilis on a medieval skeleton. Anthropologie (Brno) 30: 95-98.

Fisher T (1997) Prostitution and the Victorians. New York: Sutton Publishing.

Fournier A (1886) La Syphilis Héréditaire Tardive. Paris: G. Masson.

Gladykowska-Rzeczycka JJ and Krenz M (1995) Extensive change within a subadult skeleton from a medieval cemetery of Slaboszewo, Mogilno District, Poland. Journal of Paleopathology 7: 177-84.

Hendricks KM and Duggan C (2005) Manual of Pediatric Nutrition. Hamilton: BC Decker.

Hyde JN (1897) What Conditions Influence the Course of Infantile Syphilis? The Medical News, Dec 04 (no volume or pages numbers printed).

Kassowitz M (1882) Die Normale Ossification und die Erkrankungen des Knochensystems bei rachitis und hereditarer syphilis. 2 vols. Vienna: Wilhelm Braumüller.

Mays S, Brickley M and Ives R (2006) Skeletal manifestations of rickets in infants and young children in a historic population from England. American Journal of Physical Anthropology 129: 362-74.

Ortner DJ and Ericksen MF (1997) Bone changes in the human skull probably resulting from scurvy in infancy and childhood. International Journal of Osteoarchaeology 7: 212-20.

Ortner DJ and Mays S (1998) Dry-bone manifestations of rickets in infancy and early childhood. International Journal of Osteoarchaeology 8: 45-55.

Redmond W and Hunter H (1876) The Beautiful Baby, or Dr. Ridges Food (song). London: Hopwood and Crew.

Swan H (1900) Girls Christian Names: their History, Meaning and Association. London: Swan Sonnenchein and Co.

Taylor RW (1875) Syphilitic Lesions of the Osseous System in Infants and Young Children. New York: William Wood.

Withycombe EG (1977) The Oxford Dictionary of English Christian Names. Oxford: Clarendon Press.

Burial Rites, Death & Disease in the Late Gallinazo/Moche Period

Richard N.R. Mikulski

Centre for Human Bioarchaeology,
Museum of London,
150 London Wall,
London
EC2Y 5HN
email address for correspondence: rmikulski@museumoflondon.org.uk

Abstract

Human remains recovered from the site of Huaca Santa Clara in the Virú valley, northwest Peru, demonstrated excellent preservation and exhibited a broad spectrum of palaeopathological conditions including healed and unhealed blunt force trauma, multiple sharp force trauma, probable infectious disease, and possible achondroplasia. In addition, the remains provided evidence of ritual funerary behaviour including movement and/or removal of remains, animal sacrifice and probable human sacrifice of children or retainers. Evidence is also provided to suggest that the Gallinazo/Moche viewed the body as part of their material culture.

Keywords: Peru, sacrifice, palaeopathology, trauma, Moche, Gallinazo.

Introduction

Huaca Santa Clara is situated in the Virú river valley, approximately 50km south of the pre-Columbian Moche heartland and the modern city of Trujillo in northwest Peru. The archaeological site itself is situated on a distinct hill, Cerro Cementario, in the middle Virú valley, close to the modern town of Virú. The environment constitutes an arid, coastal desert plain.

The original settlement dates to the Early Intermediate Period and is associated with the Gallinazo culture, which flourished between 200BC and approximately AD800, with evidence for some form of contact with the succeeding Moche culture. There is also evidence for a period of later re-use of the site during the Late Intermediate Period, from between AD1000 and AD1300.

Excavations at the site recovered a total of 13 complete or articulated individual burials. Disarticulated remains recovered indicated the presence of a minimum of 11 additional individuals. All human skeletal remains recovered exhibited an exceptional state of preservation, often being recovered in association with good preservation of soft tissues, hair, nails, insect remains, textiles and other grave goods and burial artefacts.

Materials & Methods

Initially the human remains were excavated solely by the archaeologists, but osteologists supervised later exhumations. The remains were then removed to the field laboratory for cleaning and processing. While the majority of remains were in a very good state of preservation, in some cases remnants of soft tissue or residue inhibited observation. In such cases, where the remaining soft tissue had not retained its integrity, it was carefully removed and the bones cleaned.

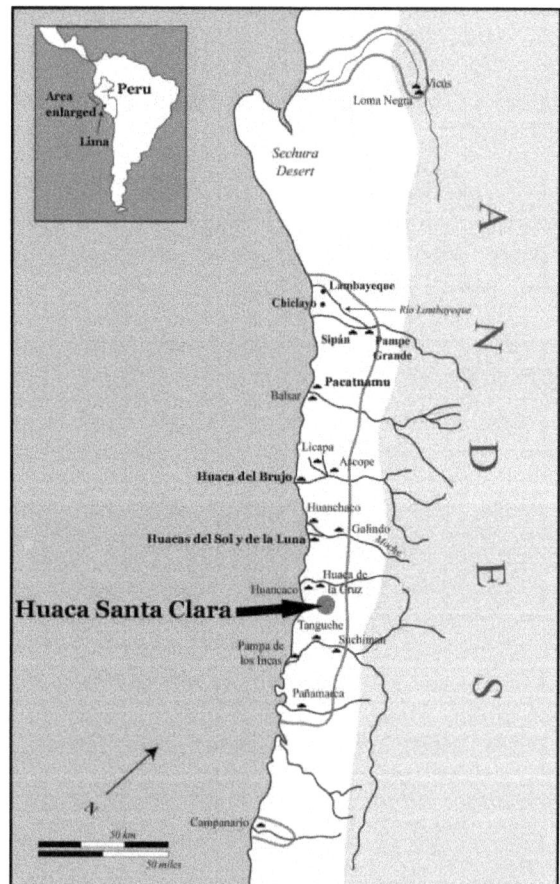

Figure 1. Map of north coastal Peru, showing major Moche centres.

Recording and analysis of the human remains was carried out following Buikstra & Ubelaker (1994), taking account of epiphyseal fusion for ageing the majority of the subadults. Adult age was generally assessed by observation of the pubic symphysis following Brooks & Suchey (1990) and the auricular surface following

Lovejoy et al. (1985). Where possible, adults were placed into young, middle or old adult age categories, following Buikstra & Ubelaker (1994). Sex was attributed to the adults (and in some cases, adolescents) by assessment of pelvic and cranial morphological traits, following Buikstra & Ubelaker (1994).

The Skeletal Material

The Early Intermediate Period (200BC – ~AD800)

Seven of the thirteen articulated burials were recovered from contexts associated with the Early Intermediate Period. All of the individuals were of adult age at the time of death, with two attributed male sex, three female sex and two of indeterminate sex (see Table 1).

Burial 7

Burial 7 consisted of the remains of a young adult female. Several textile shrouds, a textile bag and a gourd plate, in addition to the remains of a dog, were associated with the human remains. Elaborate tattoo decoration to the face, in particular around the mouth region and lower jaw, was evident

Burial 9

The individual from Burial 9 was an adult female who appeared to have been placed or 'thrown' unceremoniously into a small, shallow, unlined grave-cut. The body lay on its right side, right leg extended, left leg flexed and both arms flexed (Fig. 2). Textiles (including a possible tunic) and a gourd plate, were found with the remains.

Figure 2. Burial 9: Plan view of remains in situ.

The cranium exhibited a massive fracture to the posterior aspect, with radiating fractures and at least one other

fracture to the glabellar region with 'wet' splintering observed. These injuries were indicative of perimortem blunt force trauma and were suggestive of at least one blow to the back of the head and another to the nasal/glabellar region of the face. The woman also exhibited tattoo decoration in the form of a double line around the left wrist.

Burial 13

The remains from Burial 13 were those of an elderly female in a flexed position. She was accompanied by textiles including funerary shrouds and a wool blanket, and held a gold nugget wrapped in woollen thread in her hand. Whilst the majority of the remains were articulated, there was some indication for possible post-mortem disturbance as the coccygeal vertebrae were recovered, not from a position near to the inferior aspect of the sacrum, but from the region of the lower torso. This location seems unlikely to be explained simply as a result of collapse following post-mortem decay.

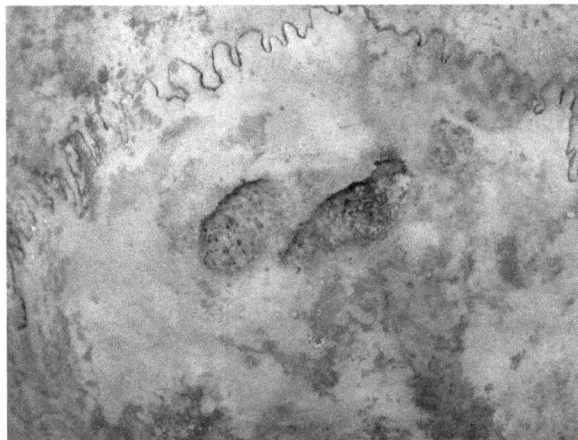

Figure 3. Burial 13: Lytic lesions to external surface of cranium.

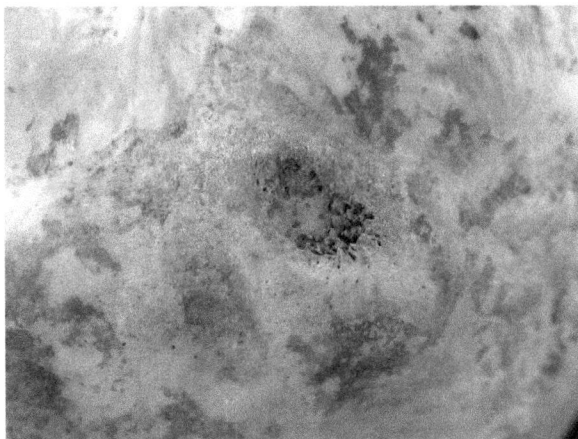

Figure 4. Burial 13: Erosive lesion to external surface of cranium, with evidence of marginal bony reaction.

Table 1. Summary of burials & associated artefacts

Burial	Period	Age	Age range	Sex	Associated Animals	Textiles	Organics	Ceramics	Other goods	Trauma
1	Early	Unaged Adult	-	Unsexed	-	-	-	yes	-	
2	Early	Unaged Adult	-	Unsexed	-	yes	yes	-	-	
7	Early	Young Adult	20 to 35	Female	Dog	yes	yes	-	textile bag	
8	Early	Unaged Adult	-	Male	-	yes	-	-	-	
9	Early	Young Adult	20 to 35	Female	-	yes	yes	-	gourd plate	Blunt force
13	Early	Old Adult	40+	Female	-	yes	-	-	gold nugget	Blunt force
14	Early	Old Adult	40+	Male	-	yes	yes	-	cane coffin, copper objects	
3	Late	Infant	~ 5	Unsexed	Llama	-	-	-	-	Sharp force
4	Late	Adolescent?	-	Female?	Llama	-	-	yes	ceramic cup	Accidental
5	Late	Infant	5 to 10	Unsexed	Llama	-	-	-	-	
6	Late	Adolescent	10 to 15	Female?	Llama(s?)	yes	yes	-	textile rope	Blunt force
10	Late	Infant	5 to 10	Unsexed	-	yes	-	-	-	Sharp force
11	Late	Adolescent	10 to 20	Unsexed	-	yes	-	-	-	

The individual exhibited five lytic-looking lesions to the outer table of the cranium. Two were separate, rounded and scooped lesions to the central squamous portion of the occipital, two lesions in the process of coalescing were in the region of the left parietal boss (again rounded and scooped), and a further lesion was located on the right parietal. There was also an additional single lesion also to the left side of the face within the superior left zygomaticomaxillary suture. While some of these cranial lesions appear to demonstrate some remodelling around their margins, others had very sharp edges indicative of a purely lytic or erosive process (Fig. 3). All the lesions appear to originate on the outer table and none of them penetrate beyond the outer layer of diploic bone. While one of the lesions may be seen to be similar to early caries sicca (Fig. 4), the lack of changes to the long bones characteristic of treponematosis, would seem to preclude this diagnosis. The most likely explanation for the cranial lesions is some type of mycotic infection (Ortner pers comm.).

The individual from burial 13 also exhibited multiple unhealed fractures, with evidence for secondary infection. The most dramatic of these was a hyperplastic fracture to the distal left tibia, with advanced callus formation and a massive cloaca indicating substantial secondary infection. Also present were a hyperplastic non-union fracture to the midshaft of the ulna, (with the unhealed ends exhibiting eburnation, indicating pseudoarthrosis) and a well-healed fracture to the right clavicle.

Burial 14

Burial 14 consisted of a cane tube coffin, apparently in-situ, containing the remains of an elderly adult male. The body lay extended supine within the coffin and was wrapped in at least two woollen blankets (preservation of which was excellent). Large, dried gourd 'bowls' were recovered from beneath the head and the feet, and are likely to have been used originally to seal the ends of the cane tube, as found elsewhere (Donnan 1995: 128-129). Copper objects had also been placed in the hands and mouth, and by the feet of the individual.

Although one end of the outer cane tube showed evidence of probable disturbance, the remains and their associated artefacts within appeared untouched by looters. The left clavicle, however, was found to be missing after removal of a large textile 'collar' which was thought to have been covering it.

The Late Intermediate Period (c. AD1000 - AD1300)

The remaining six articulated individuals recovered were derived from contexts associated with the Late Intermediate Period. This is potentially 400 years later than the earlier period of occupation. All but one of the individuals appeared to be subadults. Despite being of subadult age, female sex was tentatively attributed to two of the individuals (burials 4 and 6). Burial 4 is of particular interest (see below), as despite having unfused

epiphyses, the long bones appear well developed and hence skeletal ageing proved problematic.

Burial 3

Burial 3 contained an individual of approximately 5 years of age associated with the remains of a llama. There was evidence of unhealed sharp force trauma to the anterior torso of the individual. Several ribs exhibited straight, diagonal cuts and one right rib demonstrated 'wet' splintering at the site of an incomplete cut through the sternal rib end (Fig. 5). The sternum also appeared to have been completely bisected by a straight horizontal cut with no healing apparent (and cause by post-mortem damage deemed to be unlikely).

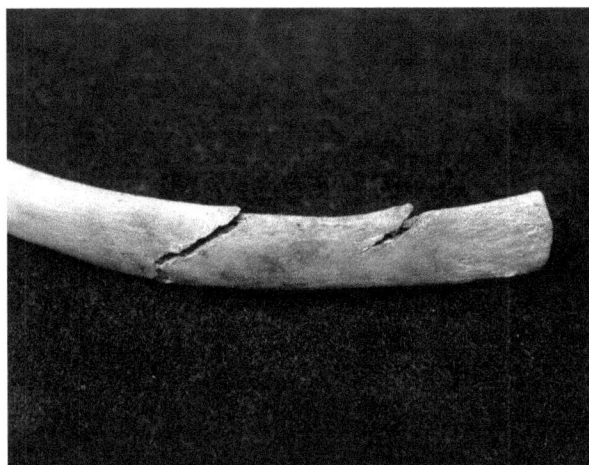

Figure 5. Burial 3: Perimortem sharp force trauma to sternal rib ends, with 'wet' splintering.

Burial 4

The incomplete, but articulated, remains from burial 4 represented an adolescent individual, of possible female sex, associated with the remains of a llama and a ceramic cup.

The development of the preserved long bones appeared unusually well advanced and very robust. Although several epiphyses remained unfused, the robustness of the arm bones seemed to indicate the individual was older than suggested by the lack of epiphyseal fusion. It is possible that this individual represents a case of achondroplasia, but this is uncertain given the absence the majority of the long bones. The right forearm also exhibited matching very well-healed fractures to the distal radius and ulna, suggestive of probable Colles fracture.

Burial 6

Burial 6 contained the remains of an adolescent of probable female sex. Associated grave goods comprised several textiles including a shawl, a gourd plate and the remains of at least one llama. The individual was found to have a textile rope around her neck (Fig. 6), possibly representing a ligature. Evidence was also found of antemortem blunt force trauma, with a circular, healed, depressed fracture located on the calotte of the cranium.

Burial 10

These remains were of a subadult individual found in association with textiles. Like the individual from burial 3, the ribs from this individual exhibited evidence of cut marks and complete cuts to the sternal rib ends. The sternum also exhibited multiple shallow cut-marks across the anterior aspect of one sternal body segment (Fig. 7). A second body segment was completely bisected and exhibited a cutmark roughly parallel to the cut section.

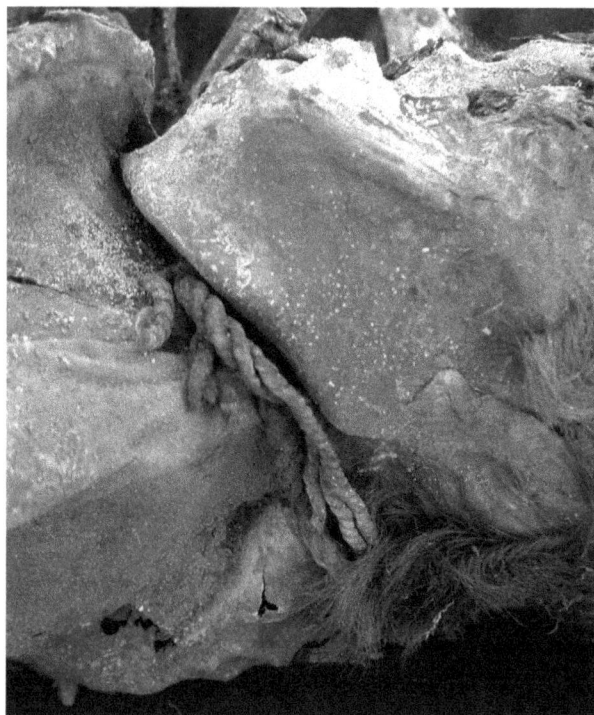

Figure 6. Burial 6: Textile rope beneath chin and around neck.

Figure 7. Burial 10: Perimortem sharp force trauma to sternum

Disarticulated Remains

Table 2. Summary of disarticulated human skeletal remains. M = Male, F = Female, ? = indeterminate.

Age	M	F	?	Unsexed
Infant	-		-	-
Adolescent	-	1?	-	4
Young adult	-		-	-
Old adult	1	1	-	-
Unaged adult	1	2	-	-
Unaged subadult	-	-	-	-
Unknown	-	1?	-	-

Amongst the disarticulated remains, there was one case suggestive of possible ankylosing spondylitis. Multiple thoracic vertebrae were fused by vertical osteophytes and several ribs of both sides were also apparently fused to the spine. There was no anterior kyphosis.

Burial Rites & Funerary Behaviour

Burial types recorded included supine cane coffins, flexed mummy bundles, subadult llama burials and one example of a shallow, 'informal' grave. In the majority of the formal burials, the individuals were found with textile garments and were wrapped in shrouds. Grave goods included a gold nugget wrapped with a woollen yarn, found in the hand of the elderly female from burial 13. The copper objects placed in the hands and mouth and by the feet of the elderly male in Burial 14 conform with a practice reported previously (Donnan & Mackey, 1978; Millaire, 2002) The cane coffin and other burial furniture of burial 14 almost exactly match other Moche burials described previously (Donnan, 1995).

Discussion

Palaeopathology

Pathological conditions observed include sharp and blunt force trauma, infectious disease, seronegative arthropathy and congenital conditions. While the remains are derived from contexts spanning a broad timescale, it is unusual to observe such a variety of different conditions within such a small sample from one archaeological site. While the trauma patterns would appear to conform to the violent social environment ascribed to the northwest coast of Peru during the time of the Moche (Benson & Cook, 2001; Bourget, 2001a; Verano, 1995; 2001a), the other pathological conditions may shed light on the environment and activity of the local population and the potential social ramifications.

It is possible that the unhealed fractures and the secondary infection described in the individual from burial 13 are related to the lesions observed on the cranium. Pathological fractures have been observed with several types of mycotic infection, including sporotrichosis and mucormycosis (Ortner 2003: 327-328;

Aufderheide & Rodriguez-Martin 1998: 221) which exhibit similar lytic lesions to the cranial vault. The apparent restriction of the lesions to the cranium, however, might argue against mycosis as a causative pathological agent. Furthermore, it is strange that there are at least two types of lesion on the cranial vault. The majority of the lesions exhibited an almost purely lytic process with sharp margins and slight internal sclerosis (figure 3), while one lesion (to the right parietal) exhibited an irregular margin with reactive periostitis evident (figure 4).

Evidence for human sacrifice

At least two of the burials from the Late Intermediate Period exhibit strong evidence for possible ritual human sacrifice. Verano (2001a; 2001b) reports that approximately 75% of individuals from Plaza 3A at Huaca de la Luna who had complete cervical spines exhibited cutmarks, with the number of cutmarks observed in each individual varying from between 1 and 9. The perimortem sharp force trauma to the chest, along with the formal disposition of burials 3 and 10 and their apparent association with burial 6, suggests they may have been ritually sacrificed at the time of inhumation of the individual from burial 6. The possibility, however, that these incisions and cuts might have been made soon after the individuals' deaths, possibly as some part of the funerary process, cannot be excluded. Equally, it is possible that the cuts may have been carried out as part of some medicinal treatment or intervention, had the individuals already been ill or diseased, although no obvious skeletal changes indicative of illness were observed.

Burial 6, whilst demonstrating no obvious signs of sharp force trauma, also exhibits possible evidence for sacrifice in the form of the observed ligature around the neck of the individual. Other examples of such ligatures have been reported at Pacatnamú by Uhle (Verano 2001b: 165) and at El Brujo (Verano 2006). Caution should be exercised in this interpretation as it might simply have formed part of the funerary treatment rather than a means of dispatching the individual.

Verano (2006) has characterised Early Intermediate period ritual human sacrifices in north coastal Peru as fitting into one of two main groups. At sites such as Huaca de la Luna and Pacatnamú, the remains of multiple groups of individuals were simply left exposed on the ground (Bourget, 2001a; Verano, 2001a; 2001b) with no formal burial treatment. These individuals appeared to be, without exception, males in the age range of 15 to 35 and exhibited evidence of traumatic deaths, including unhealed sharp and blunt force trauma, mutilation and disarticulation. The second main type of ritual human sacrifice appears to be that of retainers or children, usually associated with the burials of select individuals (Bourget, 2001b; Shimada et al. 2004; Verano, 2001a; 2006: 9). Such examples are characterised by a great deal of care and attention accorded to the individuals and their burial, with formal

positioning of the body and are associated with animal sacrifices (typically llama) and grave goods. The skeletal remains from the Late Intermediate period at Huaca Santa Clara appear to have more in common with the latter group described by Verano, with the young female from burial 6 potentially constituting the principal individual.

Evidence for unauthorised killing?

It is also possible that burial 9 was a ritual killing. This burial may be a sacrifice, associated with the tattooed female from burial 7, or a dedicatory offering (for further descriptions, including the associated architecture, see Verano 2006). It seems odd that the remains were buried in a grave without the formal treatment usually accorded to retainer sacrifices. The body was in an atypical position for a formal burial. Furthermore, grave furniture, funerary textiles and grave goods were absent. The nature of this individual's burial, combined with the perimortem trauma and the uncharacteristic 'informal' position of the body suggest the possibility of an 'unauthorised' killing.

The only comparable instance of such informal treatment of an inhumed individual is from Tomb 2 at the Moche site of Sipan (Verano 1995: 197-198). Here, a young adult female "lay sprawled face down in a position suggesting casual treatment of the body rather than careful placement" (Verano 1995: 197-198). No ligature was observed: neither were there any clear signs of skeletal trauma (Verano 2001b: 169).

Possible post-mortem movement of human remains/removal of elements

The evidence of post-mortem movement of the remains and/or removal of elements of the body (such as burial 14) suggests some sort of extended funerary process. Evidence for such extended funerary behaviour has been reported previously (Millaire 2004; Nelson 1998; Shimada et al. 2004).

The removal of the clavicle from the individual in burial 14 is intriguing, as it appeared to have been 'replaced', or its removal hidden, with a thick collar of textiles. The behaviour observed is similar to that observed in a Gallinazo burial of an adult female from Cerro Blanco, excavated by Robert Feldman in 1972 (Donnan & Mackey 1978). In that case, the skull was missing, while the rest of the body remained articulated in situ. In the position of the head, a ceramic vessel had been placed. Whether the skull had been removed prior to burial, or at some point afterward exhumed with care, the vessel appears to have been a 'replacement', especially given its painted decoration of a stylised face.

Verano (1995: 192) notes that "portions of enemies … were occasionally retained for ritual or display purposes" and describes the phenomenon of missing radii at the site of Pacatnamú. Millaire (2004) proposes the possibility of the Moche having had charnel houses and suggests that there is potential evidence within Moche ceramics and iconography for such funerary practice, but no direct evidence of such behaviour has yet been reported.

Human remains as material culture

There has been some discussion as to whether the iconography left by the Moche, and other past societies in this region, represents a symbolic expression of their culture and beliefs or a literal depiction of contemporary life (for details, see Shimada 2004). At least two of the individuals exhibited tattoos (burials 7 and 9). The tattoos to the face of the individual from burial 7 seem particularly significant. While by no means definite, these markings bear a similarity to depiction of heads from decorated ceramic vessels (Fig. 8).

Figure 8. Detail from a rollout drawing of painted decoration on a vessel in the American Museum of Natural history, New York; drawing by Donna McClelland. In Pillsbury, 2002: p.114.

Bourget (2001a: 99) has previously noted a bias in the depiction of the lower jaw/mandible in Moche iconography. He observes that this emphasis on the lower jaw is evident in the many examples of disarticulated mandibles, and in the majority of clay effigies from Huaca de la Luna, with designs painted on the jaws "extending from cheek to cheek" (Bourget 2001a: 99). There are other examples of the material culture directly reflecting the treatment of human remains. The most notable are the Moche 'trophy' heads (Verano 1995; 2001b: 173-174), where crania fashioned into drinking bowls were discovered. The sectioning of the calvarium, effectively removing the top of the cranium (calotte) is considered identical to cranium-shaped ceramic drinking vessels found.

Conclusions

Early Intermediate Period

The evidence for cultural modification of the body (specifically the tattoos observed upon the individuals from burials 7 and 9) appear to corroborate iconographic sources of cultural practices. This supports the argument

that the iconography of this period and region was a literal record of the local culture rather than a purely symbolic expression of ideas and themes.

The absence of the left clavicle from burial 14 appears to indicate at least some sort of extended funerary practice. It may suggest that the body or parts thereof had the potential to pass directly into the material culture itself.

The female from burial 7 appears to be an individual of high status, but whether her inferred high social standing was a consequence of her involvement in a ritual ceremony (perhaps ritual sacrifice) or whether she was an important figure in life who died naturally, we cannot be certain. The importance attached to her seems indisputable, given her facial tattoos, the relatively large quantity of grave goods buried with her and her association with other individuals who exhibit possible evidence of having been sacrificed (e.g. burial 9).

In summary, it seems obvious that further investigation and study of the human remains from this region will complement iconographic sources derived from the ceramic and architectural material culture.

Late Intermediate Period

The association of the burials from the Late Intermediate period, the fact that all the individuals were subadults or young adults, the careful placing of subadult llama remains within many of the burials and the evidence of probable sacrifice in at least two of the individuals suggest some form of ritual (possibly a single event or ceremony). These sacrifices of children and/or perhaps retainers demonstrate parallels with other Pre-Columbian sites, such as Tomb 2 at Sipan.

If the individual from burial 9 was sacrificed as part of a ritual, possibly associated with the inhumation of the tattooed individual from burial 7, this would lend support to the idea of long-standing cultural traditions regarding human sacrifice in this region. Burials 7 and 9, from the Early Intermediate period, and burials 3, 4, 5, 6, 10 and 11, from the Late Intermediate period, were buried within the architectural structure of the settlement. The association between architectural features and human remains (both articulated and disarticulated) has been noted previously, and the possibility of the human remains being dedicatory sacrificial offerings related to new phases of construction has been suggested (Bourget 2001a; 2001b; Shimada 2004; Verano 2006).

Evidence derived from the human remains recovered from Huaca Santa Clara indicates a high incidence of and variation in pathology. The majority of the burials from the earlier period of occupation contained individuals of adult age and several of these appear to have individuals of relatively high status (burials 7, 13 and 14). In contrast, the individuals recovered from the later period of occupation appear to have been exclusively subadults or young adult females. All of these later burials were recovered from associated contexts, and several exhibit evidence suggestive of human sacrifice. The disposition of these burials, with their apparent offerings of subadult llamas and funerary textiles suggests that this inhumation or series of inhumations related to some form of ritual ceremony.

Acknowledgements

The author thanks Jean-François Millaire (McGill University), Steve Bourget (University of Texas at Austin) and Rosa Cortez Vilchez (Université de Montréal) and two reviewers for their comments and support.

Literature Cited

Aufderheide AC and Rodriguez-Martin C (1998) The Cambridge encyclopaedia of human palaeopathology. Cambridge University Press: Cambridge, UK.

Benson EP and Cook AG (eds.) (2001) Ritual sacrifice in ancient Peru. University of Texas Press: Austin, USA.

Bourget S (2001a) Rituals of Sacrifice: Its Practice at Huaca de la Luna and Its Representation in Moche Iconography. In J Pillsbury (ed.): Moche art and archaeology in ancient Peru. New Haven: Yale University Press: 88-109.

Bourget S (2001b) Children and Ancestors: Ritual Practices at the Moche Site of Huaca de la Luna, North Coast of Peru. In EP Benson and AG Cook (eds.): Ritual sacrifice in Ancient Peru. Austin: University of Texas Press.

Brooks ST and Suchey JM (1990) Skeletal age determination based on the Os Pubis: A comparison of the Acsadi-Nemeskeri and Suchey-Brooks Methods. Human Evolution 5: 227-238.

Buikstra JE and Ubelaker DH (1994) Standards for data collection from skeletal remains. Arkansas Archaeological Survey Research Series No. 44. Fayetteville, Arkansas.

Dillehay TD (ed.) (1995) Tombs for the Living: Andean Mortuary Practices: A symposium at Dumbarton Oaks, 12th-13th October, 1991. Dumbarton Oaks Research Library and Collection: Washington D.C., USA.

Donnan CB (1995) Moche funerary practice. In TD Dillehay (ed.): Tombs for the Living: Andean Mortuary Practices: A symposium at Dumbarton Oaks, 12th-13th October, 1991. Washington D.C.: Dumbarton Oaks Research Library and Collection: 111-160.

Donnan CB and Mackey CJ (1978) Ancient Burial Patterns of the Moche valley, Peru. University of Texas: Austin, USA.

Lovejoy CO, Meindl RS, Pryzbeck TR and Mensforth RP (1985) Chronological metamorphosis of the auricular surface of the ilium: A new method for the determination of age at death. American Journal of Physical Anthropology 68: 15-28.

Millaire J (2002) Moche burial Patterns: An investigation into prehispanic social structure. British Archaeological Reports International Series 1066.

Millaire J (2004) The Manipulation of Human Remains in Moche Society: Delayed burials, grave reopening, and secondary offerings of human bones on the Peruvian north coast. Latin American Antiquity 15: 371-388.

Nelson AJ (1998) Wandering bones: Archaeology, forensic science and Moche burial practices. International Journal of Osteoarchaeology 8: 192-212.

Ortner DJ (2003) Identification of pathological lesions in human skeletal remains. Academic Press: San Diego.

Pillsbury J (ed.) (2001) Moche art and archaeology in ancient Peru. Yale University Press: New Haven.

Shimada I, Shinoda K, Farnum J, Corruccini R and Watanabe H (2004) An Integrated Analysis of Pre-Hispanic Mortuary Practices: A Middle Sican case study. Current Anthropology 45: 369-402.

Verano JW (1986) A Mass Burial of Mutilated Individuals at Pacatnamu. In CB Donnan and GA Cock (eds.): The Pacatnamu Papers, vol. 1. Los Angeles: Museum of Cultural History, University of California; 117-138.

Verano JW (1995) Where do they rest? The Treatment of Human Offerings and Trophies in Ancient Peru. In TD Dillehay (ed.): Tombs for the Living: Andean Mortuary Practices: A symposium at Dumbarton Oaks, 12th-13th October, 1991. Washington D.C.: Dumbarton Oaks Research Library and Collection; 189-227.

Verano JW (2001a) War and death in the Moche world: Osteological evidence and visual discourse. In J Pillsbury (ed.): Moche art and archaeology in ancient Peru. New Haven: Yale University Press; 111-125.

Verano JW (2001b) The Physical evidence of human sacrifice in ancient Peru. In EP Benson and AG Cook (eds.): Ritual Sacrifice in Ancient Peru. Austin: University of Texas Press.

Verano JW (2006) Human Sacrifice at El Brujo, Northern Peru: Report on 2005 Summer field research supported by the National Geographic Society's Committee for Research and Exploration (Grant #7844-05) and the Roger Thayer Stone Center for Latin American Studies (Faculty Summer Research Grant).

Overweight and the Human Skeleton

Philippa Patrick

Department of Haematology,
Addenbrooke's Hospital,
Hills Road,
Cambridge CB2 2QQ
philippa.patrick@addenbrookes.nhs.uk

Abstract

This paper reviews the scope for identifying overweight individuals in archaeological skeletal material, which has received relatively little attention in the osteoarchaeological literature to date. Obesity-related joint disease is briefly discussed, but the chief focus of this paper is the testing of techniques proposed by McHenry, Ruff and Porter for extrapolation body weight and body mass index, both in terms of their general usefulness and their scope for identifying over- and underweight individuals in archaeological skeletal material. Particular concerns are raised regarding the use of techniques developed on the basis of living individuals and the use of radiographic imaging.

Introduction

The effect of being overweight on the human skeleton has received relatively little attention in the archaeological and biological anthropological literature to date and the issue of overindulgence in the past is far outweighed by concentration on malnutrition (e.g. Larsen 1997). There could be a number of reasons for this discrepancy, including the confinement of obesity-related pathologies to older individuals and the fact that the aetiology of these diseases is complex; one condition being relatively little understood even in the clinical literature and another having a multifactorial aetiology. There is also the potential for author bias and the lack of attention to skeletal markers of overweight may to an extent reflect a misconception that past lifestyles did not present the opportunity for becoming overweight (Patrick 2005a: 98). Although as Garrow (1992:60) notes, obesity may be regarded as "a natural consequence of affluence and longevity", those factors are not exclusive to the modern day and the opportunity for overindulgence would have existed in the past. Physiological and especially genetic factors regarding how the body stores and distributes fat also have an important part to play, (Campbell 1998:1137; Ferrell 1993), interacting with more 'social' or 'behavioural' factors and tending toward obesity.

This paper explores the potential for identifying the overweight and its effects upon archaeological human remains, from the point of view of both pathology and biomechanics. It focuses in particular on the strengths and weaknesses of various approaches to weight and body mass index extrapolation, based on the outcomes of a study of 274 males over the age of 45 from medieval cemetery sites in Greater London (Patrick 2005b). The age range of the sample reflects the fact that older individuals tend to develop the joint diseases that are associated in the clinical literature with obesity, while the sex distribution is significant only insofar as it reflects an initial research question pertaining to medieval monastic lifestyle. The assemblage presented in this paper incorporates both monks and laymen, and dates from approximately AD 1066-1540.

Overweight, and its more extreme manifestation, obesity, can be considered both as a cultural and a biological phenomenon. Ulijaszek (1998:410) provides a useful summary of this distinction, stating that "a social definition of obesity is one of fatness beyond the socially accepted norms of a given society, while a medical definition relates to individuals who weigh more than the upper acceptable limit for their height and frame". It is the latter, medical concept of overweight that osteoarchaeologists can seek to explore. As Ulijaszek pointed out, height is an important consideration as well as weight when exploring questions of physique, and this paper will explore the potential for applying a weight-for-height index, namely the body mass index (BMI; weight/height^2) to skeletal remains. Guidelines developed by the World Health Organisation (WHO) state that a BMI of less than 20kg/m^2 is defined as "lean", a BMI of 20-24.9 kg/m^2 is "acceptable", a BMI of 25-29.9 kg/m^2 is "moderately overweight", a BMI of 30-34.9 is "severely obese" and a BMI greater than 35 is "morbidly obese" (Lieberman 2000:1066).

Materials and methods

Obesity-related pathology

Overweight affects the body in a number of ways, increasing the risk of developing type II diabetes, cardiovascular disease and pulmonary disorders (Passmore & Eastwood 1986:274). In terms of skeletal pathology, obesity is cited as an aetiological factor underlying certain forms of osteoarthritis (Spector 1990:283), and an association has also been suggested with the still relatively poorly-understood condition diffuse idiopathic skeletal hyperostosis, possibly related to type II diabetes (Julkunen et al. 1971:605).

The three forms of osteoarthritis (OA) particularly associated with obesity in the clinical literature are OA of the knee, the hip and the distal interphalangeal joints of the hands, while possible links have also been suggested with OA in the lumbar spine, feet and sternoclavicular joint of the shoulder (Hartz et al. 1986:311). The link between obesity and OA of the knee has wide support in the clinical literature (e.g. Coggon et al. 2001; Leach et al. 1973), while there has been considerably more debate surrounding the link between obesity and OA of the hip (e.g. Saville & Dickson 1968; Lievense et al. 2002). The strong link between obesity and OA of the distal interphalangeal joints of the fingers demonstrates that the relationship with the disease is more complex than a straightforward issue of greater body mass causing stress to weight-bearing joints; a metabolic explanation for this association has been suggested (Hartz et al. 1986: 311; Cooper & Dieppe 1994:23). Alteration of joint mechanics because of increased depositions of adipose tissue have been cited in the development and distribution of knee osteoarthritis, with Radin et al. (1972) suggesting that individuals with excessive fat on their thighs alter their gait, walking in a slightly 'bow-legged' fashion, which thus puts excessive pressure on the medial tibio-femoral joint.

Osteoarthritis has a multifactorial aetiology, thus obesity interacts with an array of other 'susceptibility' and 'mechanical' variables such as genetic predisposition to OA, joint shape and repeated activity, to result in joint failure (Dieppe 1991:9; Cooper & Dieppe 1994:22-27). Where other predisposing factors are not present, obese individuals do not necessarily go on to develop osteoarthritis. OA predominantly affects older individuals, and so cannot be used as a reliable indicator of overweight in individuals under the age of 45.

Diffuse idiopathic skeletal hyperostosis (DISH; Forestier's disease) is a phenomenon first described in the clinical literature in 1950, and is symptomatic only as "a stiffness of the spine" (Forestier & Rotés-Querol 1950:322). In the clinical literature, DISH is largely discussed in the context of radiographic studies (e.g. Resnick et al. 1975), yet its characteristic lesion (florid, flowing ossification of the anterior longitudinal ligament of the spine (Resnick & Niwayama 1976:559), involving only one side of the thoracic spine (typically the right)), facilitates diagnosis on visual analysis of skeletal remains (Rogers et al. 1997:85). The hip (in particular fusion of the sacro-iliac joint), the foot and the shoulder may also be involved (Rotés-Querol 1997:1194). As with osteoarthritis, DISH is a disease that tends to affect older people.

DISH has a considerable antiquity, yet in the archaeological literature to date, it has been particularly explored in relation to medieval monasticism (Waldron 1985; Rogers & Waldron 2001; Rogers 2000:171; Patrick 2005b). Monastic lifestyle, characterised by little physical activity and calorific intake well in excess of modern recommended daily intakes (Harvey 1993:70) epitomises the supposed aetiology of DISH, and high prevalences are consistently observed in monastic contexts; however there is ample scope for further investigation with regard to other periods and populations.

Body weight extrapolation

Aside from pathology, it is also possible to explore overweight in terms of biomechanics and skeletal remodelling in order to accommodate increased load-bearing stress due to overweight in line with Wolff's law of bone remodelling (1892). Most extant techniques for body mass extrapolation were developed by physical anthropologists. Although there are some forensic applications of the methods, the more recent nature of forensic material means that other indicators of physique are available, such as the clothing size (Morse et al. 1983:106).

The chief aim of weight extrapolation in physical anthropology is to determine the likely body weight of early hominid specimens and a wide variety of methods have been proposed for this purpose. When selecting which techniques to use, it is necessary to consider the biomechanics of the species in question. In a bipedal species such as *Homo sapiens*, weight is transmitted down the spine and legs. Therefore in terms of load-bearing function, the femora, tibiae and lumbar spine should yield the most useful measurements for weight extrapolation. All these elements tend to preserve fairly well archaeologically (Waldron 1987:58-60; 62).

This paper discusses the weight extrapolation techniques of three researchers; Henry McHenry, Christopher Ruff and Alan Porter. Other methods such as the skeletal weight-based technique proposed by Steudel (1980) were excluded, as they require many skeletal elements to be preserved: this is rarely possible in the light of variable archaeological preservation. All the techniques discussed here are non-destructive, utilising either external bone measurements or radiographic imaging. Although destructive techniques can give more accurate measurements, especially when considering cortical areas, it is necessary to consider whether the advantages of such techniques justify the potential data loss to future researchers. Figure 1 summarises the measurements used for weight and height extrapolation, and Table 1 presents the regression equations.

Table 1. Regression equations for the extrapolation of body weight in anatomically modern humans. All extrapolated weights in kg

Measurement	Reference	Equation
Femoral shaft cortical area	Ruff et al 1991	wt = 0.0808 x CA + 39.4
Femoral head diameter	Ruff et al 1991	wt = 3.383 x H - 85.8
TV12 waist area	McHenry 1992	log wt = 0.6552 x log T12 - 0.2443
LV5 waist area	McHenry 1992	log wt = 1.1593 x log L5 - 1.9630
Femoral head diameter	McHenry 1992	log wt = 1.7125 x log femhead - 1.0480
Femoral shaft area	McHenry 1992	log wt = 0.7927 x log femshft - 0.5233
Femoral shaft area	McHenry 1988	log wt = 0.624 x log femshft - 0.0562
Proximal tibia area	McHenry 1992	log wt = 1.0585 x log proxtib - 1.9537
Proximal tibia area	McHenry 1991	log wt = 1.0583 x log proxtib - 1.9537
TV12, LV5 superior & waist areas	McHenry 1975	log wt = 0.6031 x vertebrae + 1.085
LV1 waist area	Porter 1997	wt = 4.66 x lumare1 + 8.54
Mid-shaft tibia A-P	Porter 1997	wt = 17.8 x midwid + 4.45
Femur length	Porter 1997	100/wt = -0.031 x femlengh + 2.82

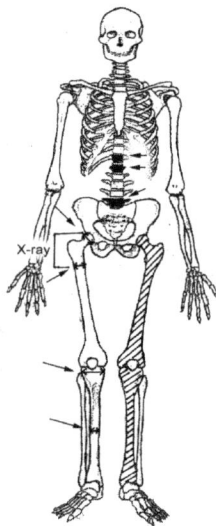

Figure 1. Bone and x-ray measurements discussed in this paper

Nine regressions proposed by McHenry (1975; 1988; 1991; 1992) were used, based on vertebral and lower limb measurements. Three regressions suggested by Porter (1997) were selected, in each case using the extrapolation for white males, as this best reflects the medieval English population, and the fact that all individuals studied were male. Again the selected techniques were based on the lumbar vertebrae and lower limbs. One technique proposed by Ruff et al. (1991) was selected, utilising x-rays of the proximal femur. They tested a number of variables using medical x-rays, with a view to establishing a reliable indicator of both current weight (which would in theory equate with weight at death in skeletal collections), as well as recalled weight at 18, allowing investigation of weight change during adulthood. They recommend using femoral head diameter as a proxy for weight at age 18, and femoral shaft cortical area for current weight. The study of cortical thickness has the advantage that it may give

insights into both external and internal buttressing of cortical bone, in accordance with Wolff's Law of bone remodelling.

Each researcher based regression equations on very different data sets. Only Porter proposed regressions specific to sex and race, making it possible to select an equation tailored to a particular target population. His use of living, healthy individuals meant that current body weight could be accurately recorded. Like Porter, Ruff et al. (1991) used only anatomically modern humans. The technique is based on living, otherwise healthy individuals and again the current body weight could be accurately known. Recalled weight at 18 was used – relying on accurate and honest reporting by subjects. McHenry's techniques were largely based on cadaveric reference collections such as the Terry Collection, comprising North Americans of mixed ancestry, along with pygmies, Khoisan and specimens of other hominin species, reflecting his focus on early hominid body weight rather than anatomically modern humans, which have large skeletons compared to early hominids (McHenry 1992:408). Potential concerns arise as to whether the weights recorded at autopsy are representative of optimal healthy weight in life (Porter 1997:31), and the extent to which the combination of different races and species affects applicability to one particular population, Caucasian male anatomically modern humans.

Body mass index extrapolation

There are two potential approaches to calculation of body mass index on the basis of skeletal material. The first is to use an extrapolated weight and an extrapolated height to calculate a body mass index. For the purposes of this paper, stature was estimated using Trotter's stature regression based on femoral length (1970: 77), unless otherwise stated.

Table 2. Regression equations for extrapolation of body mass index in anatomically modern humans. All extrapolated BMIs are in kg/m^2. LV1 area/tibia length constant has been corrected from an originally stated constant of +40.

Measurement	Reference	Equation
LV1 area	Porter 1997	BMI = 0.766 x lumarea1 + 13.0
Mid-shaft tibia A-P diameter	Porter 1997	100/BMI = -0.761 x midwid + 7.36
LV1 area / tibia length	Porter 1997	BMI = 9.11 x (lumarea1/tiblength) + 20

The second means of estimating body mass index is to use a regression equation specifically aimed at extrapolation of BMI. There are relatively few such equations, but some have been proposed by Porter (1997: 73-74) using one or more skeletal measurements. The three most appropriate in terms of minimal error and likelihood of elements being sufficiently well preserved were selected for use in this study. Table 2 summarises the measurements and regression equations used

The sample and analytical methods

Two hundred and seventy-four skeletons of males aged over 45 were studied, derived from three medieval monastic sites within Greater London (Fig. 2): Bermondsey Abbey (30 individuals), Merton Priory (148 individuals), and the Royal Mint Site (96 individuals), which incorporates burials associated with St Mary Graces Abbey, Tower Hill, and burials predating the monastery's foundation, probably the Black Death cemetery of Holy Trinity East Smithfield (Grainger & Hawkins 1988). Individuals were assigned a probable monastic or lay status, based on the position of their burial in relation to the monastic church, but for the purposes of this paper, which focuses on general issues regarding physique extrapolation, the monastic/lay distinction is disregarded and the assemblages are combined.

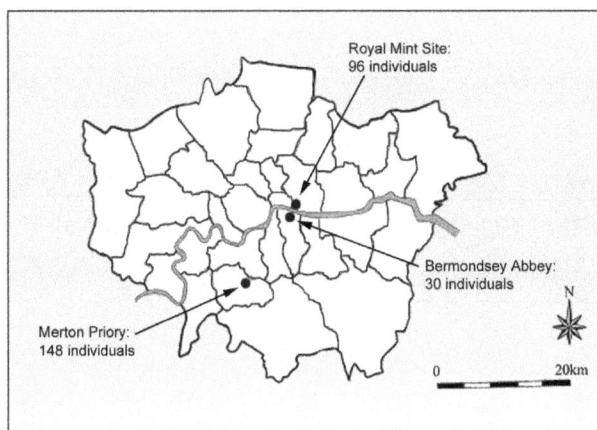

Figure 2. Map of Greater London showing the location of the three sites from which assemblages were derived.

Skeletons were inspected to determine the presence or absence of osteoarthritis and DISH. Osteoarthritis was diagnosed on the basis of eburnation of the joint surface, as advocated by Waldron & Rogers (1991). The diagnostic criterion for presence of DISH was single-sided fusion of a minimum of four adjacent thoracic vertebrae, as recommended by Rogers (2000:170). These signs are considered to be unequivocally diagnostic of their respective conditions.

External bone measurements were taken from all individuals for stature and weight extrapolation. The proximal femora of 103 individuals were imaged in a Todd Research x-ray inspection cabinet. Twelve of these specimens were also imaged with a Scanco Medical X-treme CT scanner. The latter allowed application of regression equations proposed by Ruff et al. (1991) to actual rather than approximated cortical area, and clarification of the degree of distortion produced by an x-ray cabinet. Measurements were obtained using digital callipers, with the exception of bone lengths, where an osteometric board was used and CT scans, where measurements were determined using computer software.

The usefulness of weight and BMI extrapolation techniques was assessed on the basis of average extrapolated values, and the range of values represented. A hypothetical 'ideal' technique should give median value in a 'normal' range, with a broad range of values either side indicative of the ability to isolate over- and underweight individuals. The degree of agreement between techniques was also explored; in archaeological contexts where preservation is often variable, it is useful to know whether a consensus exists between techniques using a variety of different anatomical elements, and thus whether these could be used interchangeably according to the elements adequately preserved.

Results and discussion

Average body weight

Figure 3 and Table 3 summarise the median and range of weights for each technique used. Square markers indicate techniques for which the median weight is below 70kg (11 stone), and triangular markers indicate techniques for which the median weight exceeds 70kg. This represents an arbitrary boundary only, but if techniques predicted weights significantly higher or lower than this boundary, this could highlight cause for concern regarding the technique's accuracy or applicability to skeletal material.

Two of Porter's techniques produced median weights below 70kg. Indeed with the extrapolation based on mid-shaft tibia diameter, none of the 102 individuals studied had extrapolated weights in excess of 70kg, the highest extrapolated weight being 68.19kg (10.7 stone). The lowest extrapolated body weight, produced using Porter's first lumbar vertebra regression, was 47.3kg (7.45 stone) which seems unlikely to be an accurate reflection of adult male body weight. Porter's femoral length regression gives a higher range of weights, and an average in excess of 70kg, the use of femoral length for extrapolation of body weight is of limited value since it presupposes a directly proportional relationship between height and weight. This is further explored and critiqued in discussion of body mass index below.

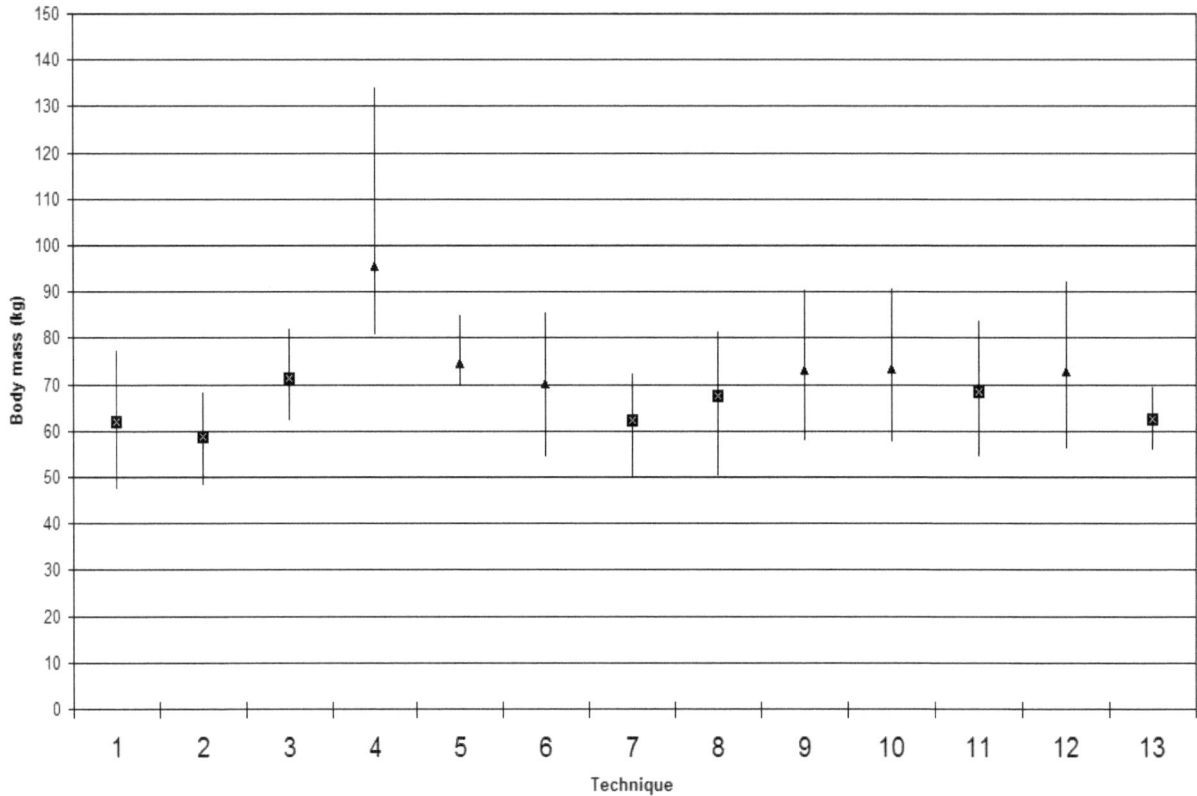

Figure 3. Comparison of median and range of extrapolated body weights. 1=Porter LV1 (n=73), 2=Porter MSTib A-P (n=102), 3=Porter Femlength (n=121), 4=Ruff CA (x-ray; n=104), 5=Ruff CA (CT; n=12), 6=McHenry 1992 Femhead (n=139), 7=McHenry 1988 Femshaft (n=219), 8=McHenry 1992 Femshaft (n=219), 9=McHenry 1991 Proxtib (n=59), 10=McHenry 1992 Proxtib (n=59), 11=McHenry 1992 T12 (n=55), 12=McHenry 1992 L5 (n=58), 13=McHenry 1975 T12&L5 (n=30).

Table 3. Comparison of median and range of body weights extrapolated using each technique (all weights in kg)

Technique	Minimum weight (kg)	Maximum weight (kg)	Median weight (kg)
Porter LV1 n=73	47.3	77.2	62.0
Porter MSTib A-P n=102	48.2	68.2	58.8
Porter Femlength n=121	62.3	81.9	71.3
Ruff CA (x-ray) n=104	80.8	133.9	95.7
Ruff CA (CT) n=13	70.0	84.7	74.5
McHenry Femhead 92 n=139	54.4	85.4	70.1
McHenry Femshaft 88 n=219	49.6	72.3	62.4
McHenry Femshaft 92 n=219	50.3	81.2	67.4
McHenry Proxtib 91 n=59	57.8	90.3	73.2
McHenry Proxtib 92 n=59	57.6	90.5	73.3
McHenry T12 92 n=55	54.5	83.7	68.5
McHenry L5 92 n=58	56.2	92.0	72.9
McHenry Vertebrae 75 n=30	55.8	69.3	62.6

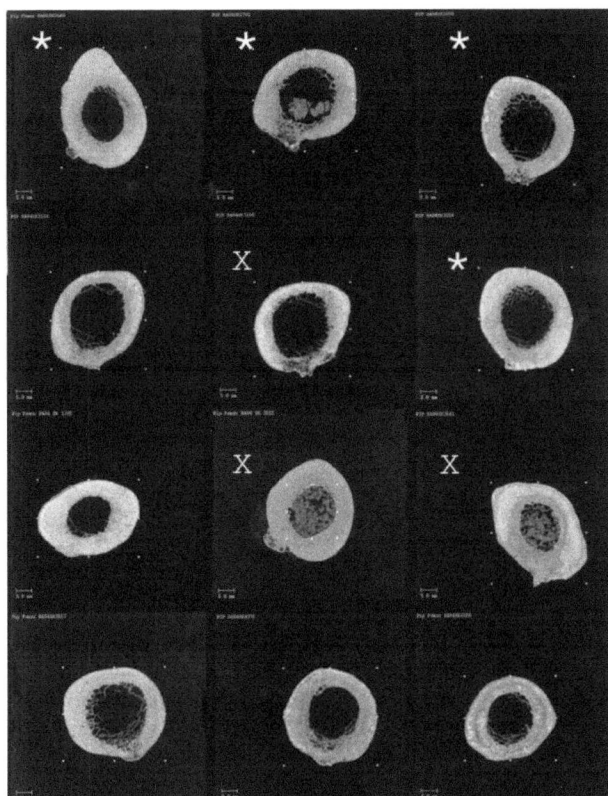

Figure 4. CT scans of femora showing variation in thickness and shape of the femoral shaft. Stars indicate individuals with obesity related joint disease. Crosses indicate individuals with no obesity related joint disease.

The tendency of Porter's techniques to provide low body weights is most likely to be related to the fact that they were developed on the basis of measurements from live individuals. The antero-posterior diameter of the tibial mid-shaft is particularly problematic, with its low range and median value (58.84kg; 9.3 stone), and it is possible that, despite using skinfold callipers to estimate the quantity of subcutaneous fat and other tissue (Porter 1997:32) the thickness of soft tissue over the bone was not corrected for accurately.

By contrast, the radiographic technique (Ruff et al. 1991) produces extremely high average body weights, in particular where plain x-rays were used as the basis for calculation of cortical area. Where CT scanning software was used to determine an exact cortical area, weights tended to be lower. Regardless of how cortical area was established, no individuals were identified as weighing less than 70kg the method, and the maximum extrapolated body weight with the x-ray technique was extremely high at 133.8kg (in excess of 21 stone). The average weight produced using calculated cortical area from x-rays was 95.66kg (15.1 stone). In the small CT sample (n=12), the median weight was 74.54kg (11.7 stone).

The problems encountered with the radiographic technique may be predominantly attributed to the

significant yet inconsistent degree of distortion on x-ray images. Furthermore, the calculated approximation of cortical area is not necessarily representative of actual area; CT scans of a sample of femora (Fig. 4) illustrate that the shape of the femur varies, and cortical thickness is rarely uniform. Table 4 compares cortical area determined using computer software, and that calculated on the basis of diameters measured from the CT scan and x-ray images. By comparing the x-ray cortical area and the CT scan cortical area, it is possible to explore the extent of distortion on x-rays. The greatest discrepancy is an exaggeration of 95% - almost double the actual area, the smallest discrepancy is 21% and the mean 51.25%. By comparing the CT scan cortical area with the cortical area approximated from diameters measured on the scan, it is possible to explore further the problems of estimating cortical area and assuming it conforms to a uniform ring-shape. The smallest difference is only 2%, while the maximum discrepancy is 78%. The mean discrepancy is 30%. In practice, the two factors (distortion and inaccurate approximation of cortical area) combine to produce overestimates of weight in the x-rayed sample.

Table 4. Comparison of plain x-ray calculated cortical area, CT scan cortical area and CT scan calculated cortical area

Skeleton	X-ray CA (mm^2)	CT scan CA (mm^2)	CT calculated CA (mm^2)
2640	784.9	560.5	597.3
2701	754.3	504.9	626.8
2898B	678.8	424.4	551.7
3104	749.3	397.3	536.7
3190	737.9	379.2	673.6
3214	746.4	488.8	621.8
3385	537.3	434.9	519.5
3501	623.2	472.3	482.4
3541B	661.2	547.0	719.5
3817	692.7	432.8	681.7
4035	634.6	422.7	517.9
4056	544.4	382.6	489.2

Despite initial concerns that McHenry's use of cadaveric collections where many individuals had died of chronic, wasting diseases might produce underestimates of body weight, this expectation is not borne out by many of his techniques, in particular his most recent regressions (1992). McHenry's techniques tend to represent a wide range of weights and the average weights for three of the techniques: the femoral head, proximal tibia and fifth lumbar vertebra, are in excess of 70kg. The technique involving the twelfth thoracic and fifth lumbar vertebrae combined, gave weights in a comparatively low range. However, the sample for this technique was small (n=30) since it required good preservation of two elements rather than just one as required by other techniques, so it would be unwise to read too much into this trend. The femoral shaft area also gives a low average weight but a fairly wide range. This is based on a much larger sample (n=219) so the low median weight is likely to be more a reflection of a possible problem with the weight regression itself than an artefact of sample size. This, in the context of Wolff's observation that long-bone shafts buttress internally *and* externally in response to stress, suggests that the external area of the femoral shaft, although often well preserved archaeologically, may not be a particularly accurate indicator of body weight.

Variation within the same individual

One of the most striking observations arising from this study was the degree of variation in extrapolated weights based on measurements for the same individual. Figure 5 shows the range of extrapolated weights from one particular individual. The minimum weight is that based on the mid-shaft tibia (Porter 1997), at 56.2kg, while the maximum is calculated from the femoral shaft cortical area (Ruff et al. 1991), at 95.6kg. It was possible to obtain all measurements and compare extrapolations for thirteen individuals (4.7% of the study sample). A consistent pattern emerged, whereby the cortical area produced the greatest weight for each individual, and Porter's techniques (predominantly the mid-shaft tibia, but in two cases the first lumbar vertebra) produced the lowest extrapolated body weight. The difference between the minimum and maximum extrapolated weight for a single individual was on average 40.1kg, although this difference ranged between 30.1kg and 56.8kg.

BA84 SK3501 - Weight comparison

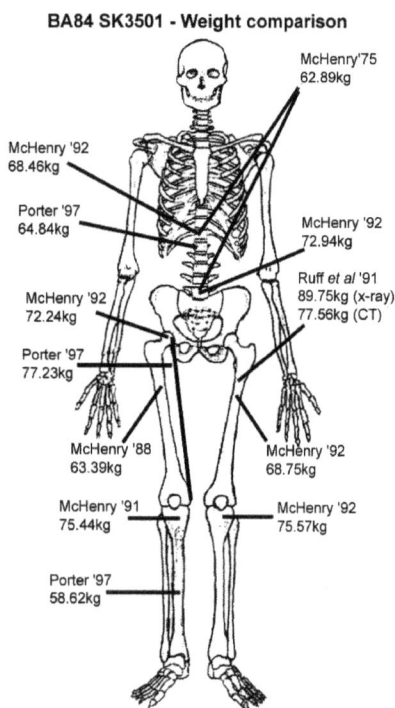

Figure 5. Comparison of extrapolated body weights for a single individual.

This degree of variation raises important considerations. Partly this reflects particular techniques producing excessively high or low average body weights, as discussed above, but importantly, it suggests that it would be imprudent to use any one of an array of possible bone measurements to calculate an individual's weight, even if this might seem the easiest option in cases of variable preservation of archaeological skeletal assemblages.

Table 5. Mean values for weight proxies in individuals with and without obesity-related joint disease.

Weight proxy	Joint disease present	Joint disease absent
Femoral head diameter	49mm (n=30)	49mm (n=44)
Femoral shaft cortical area	767mm^2 (n=15)	700mm^2 (n=31)
Femoral shaft external area	952mm^2 (n=75)	909mm^2 (n=75)
Proximal tibia area	4115mm^2 (n=17)	4062mm^2 (n=31)
Mid-shaft tibia A-P diameter	31mm (n=20)	30mm (n=40)
TV12 superior area	1563mm^2 (n=9)	1485mm^2 (n=18)
TV12 waist area	1157mm^2 (n=11)	1133mm^2 (n=19)
LV1 waist area	1217mm^2 (n=13)	1205mm^2 (n=22)
LV5 superior area	2062mm^2 (n=9)	2015mm^2 (n=23)
LV5 waist area	1544mm^2 (n=11)	1550mm^2 (n=24)

Proxies for body weight and obesity-related pathology

The usefulness of skeletal measurements as proxies for body weight can be further assessed by exploring whether there is a correlation between greater values and presence of obesity-related joint disease. Table 5 summarises the mean values for measurements used as body weight proxies and illustrates that, in the majority of cases, individuals manifesting with obesity-related OA and/or DISH tend to also have larger bone measurements. There are two exceptions to this trend; the waist area of the fifth lumbar vertebra, in which individuals without obesity-related pathology have a greater average value, and femoral head diameter, where there is no difference between the group with obesity-related joint disease, and that without. The significance of this, however, is unclear.

Femoral shaft CT scans provide scope for exploring the morphology and structure of bone and its relationship to obesity-related disease. Four of the femora that were CT scanned (Fig. 4) were from individuals with obesity-related pathology (indicated with a star) and three were from individuals found to have no obesity-related pathology (indicated with a cross). The remaining five femora were from individuals for whom there was insufficient information to exclude all forms of obesity-related pathology. No clear pattern emerges in terms of cortical area and disease status: some individuals with obesity-related pathology have very robust, thick cortical bone, yet others with no proven obesity-related joint disease have a very similar appearance. Others without joint disease, however, do have much thinner cortical bone, and, given that this sample is extremely small, and

Table 6. Comparison of median and range of body mass indices extrapolated using each technique (all in kg/m^2)

Technique	Minimum BMI (kg/m^2)	Maximum BMI (kg/m^2)	Median BMI (kg/m^2)
Ruff XrayCA/fem n=86	27.6	43.4	33.1
Ruff calcCTCA/fem n=9	24.6	33.6	28.9
Ruff CTCA/fem n=9	23.5	28.2	25.2
HcHenry FS88/fem n=96	17.2	24.8	21.9
McHenry FS92/fem n=117	17.4	27.6	23.6
McHenry FH/fem n=117	20.8	29.3	24.5
McHenry PT91/fem n=57	22.5	29.8	25.0
McHenry PT92/fem n= 57	22.6	29.9	25.1
McHenry T12/fem n=39	20.5	28.2	23.2
McHenry L5/fem n=43	19.5	31.8	24.9
Porter TibA-P/fem n=85	17.5	24.8	20.6
Porter LV1/fem n=49	17.8	26.7	22.1
Porter feml/fem n=121	24.1	26.0	24.6
Porter LV1 n=73	19.4	24.3	22.1
Porter Tibia A-P n=102	18.2	21.6	19.9
Porter LV1/Tiblength n=37	42.3	43.9	43.0
Porter LV1/Tiblength (corrected) n=37	22.3	23.9	23.0

given that not all overweight people go on to develop joint disease, the morphology of the femoral shaft would be an interesting avenue for further investigation in a group of individuals of known weight.

Body mass index

Table 6 summarises the median and range of BMIs extrapolated using each of the techniques used in this study. Bearing in mind that the WHO definition of 'normal' body weight is 20-24.9kg/m^2, we should expect the median to fall into the 'normal' category, but the range to represent both underweight (under 20kg/m^2) and overweight (over 25kg/m^2). As with weights, a reasonable range of BMIs should be represented by each technique.

This expected pattern tends to be borne out by BMIs calculated using McHenry's weight extrapolations, with medians in the normal range or slightly into the overweight range, and a wide range of BMIs represented, although the obese range is rarely entered and only a few individuals fall into the lean/underweight range. Meanwhile, as expected on the basis of observations on body weight, Porter's techniques tend to produce low BMIs (with the exception of the lumbar vertebra 1/tibia length uncorrected, where high BMIs are an artefact of an erroneous equation), and the techniques of Ruff et al. tend to produce high BMIs, extending well into the 'obese' category (over 30kg/m^2). In the case of the x-ray derived weights, all individuals fall into the 'overweight' or 'obese' categories.

BMI calculated using Porter's femoral length weight regression and Trotter's femoral length stature equation is particularly problematic. In a sample of 121 individuals, all BMIs fell into the range 24.1-26.0kg/m^2. The lack of variation in BMI suggests a directly proportional relationship between weight and height, which is to be expected given the two variables are calculated on the

basis of the same measurement. Interestingly, as Figure 7 indicates, individuals with shorter femora tended to yield slightly higher BMIs using this method. Porter's techniques for directly extrapolating BMI tend to produce BMIs within a relatively small range, and, notably in the case of lumbar vertebra 1 and tibia length, all 37 extrapolated BMIs fall within the range 22.3-23.9kg/m^2.

With their small ranges, and low median values, Porter's techniques for direct extrapolation of BMI seem unconvincing when applied to an archaeological assemblage. This is disappointing, as a single regression is clearly preferable to using an extrapolated weight and extrapolated stature, each with its own degree of error associated.

Conclusions

This paper has discussed the means of identifying the overweight from the human skeleton presently available to human osteoarchaeologists. It represents an area comparatively little-explored to date, in particular in terms of biomechanics. Furthermore little systematic study has been undertaken into the skeletal pathology associated with being overweight, especially in the context of archaeological research. This paper has highlighted the problems inherent in the study of the overweight, some of which may be remedied, or at least improved upon through future research, but some of which may be insurmountable (in particular the nature of obesity-related joint disease being such that it cannot be used to identify potentially overweight people in any but the oldest age group).

This paper has tested various regression equations for the extrapolation of body weight and body mass index using a sample of 274 archaeological skeletons, and critiqued the applicability of techniques developed on the basis of living individuals, cadaveric reference collections and radiographic material to archaeological skeletal remains. The results demonstrate little agreement between techniques and the foundation of various methods gives rise to some extremely unlikely results. They highlight the distortion effect of x-rays, as testified to by comparison with CT scan images, resulting in overestimation of body weight. Conversely, the technique developed on the basis of external bone measurements with *in situ* overlying soft tissue led to underestimation of body weight.

Future research may improve on the techniques currently available, but for the present, regressions aimed at reconstructing physique should be used with caution, bearing in mind the nature of the sample on which they are based and any resultant potential for error. For the present, until more reliable physique reconstruction techniques are available, the scope for identifying potentially overweight people in the skeletal record remains confined to older individuals where a combination of pathology and robust skeletal morphology may be considered indicative, but not necessarily conclusive evidence, of their overweight status.

Acknowledgements

This paper presents research carried out at the Institute of Archaeology, University College London, and funded by the Arts and Humanities Research Board (Grant 01/3072), with additional financial support from the Museum of London. All skeletal material used in this study was analysed by kind permission of the Museum of London's Centre for Human Bioarchaeology. CT scanning was carried out at Norwich Community Hospital with the assistance of Nicola Dalzell (Cambridge University ADOQ project) and Andres Laib (Scanco Medical). Thanks to Tony Waldron, Gustav Milne, Stuart Laidlaw, Sandra Bond (Institute of Archaeology), John Shepherd, Steve Clark and Alan Pipe (Museum of London) for their assistance and guidance during this research, to Henry McHenry, Alan Porter and Christopher Ruff for discussion of their techniques, and to Dan Hodson for helpful comments during the writing of this paper.

Literature Cited

Campbell LV (1998) A change of paradigm: obesity is not due to either 'excess' energy intake or 'inadequate' energy expenditure. International Journal of Obesity 22: 1137.

Coggon D, Reading I, Croft P, McLaren M, Barrett D and Cooper C (2001) Knee osteoarthritis and obesity. International Journal of Obesity 25: 622-627.

Cooper C & Dieppe P (1994) The epidemiology of osteoarthritis. In Doherty M (ed.): Color atlas and text of osteoarthritis. London: Wolfe: 15-28.

Dieppe P (1991) Osteoarthritis: the scale and scope of the clinical problem. In RGG Russell and PA Dieppe (eds.): Osteoarthritis: current research and prospects for pharmacological intervention. London: IBC Technical Services: 4-23.

Ferrell RE (1993) Obesity: choosing genetic approaches from a mixed menu. Human Biology 65: 967-975.

Forestier J and Rotés-Querol J (1950) Senile ankylosing hyperostosis of the spine. Annals of the Rheumatic Diseases 9: 321-330.

Garrow JS (1992) Obesity in man. In EM Widdowson and JC Mathers (eds.): The contribution of nutrition to human and animal health. Cambridge: Cambridge University Press: 53-62.

Grainger I and Hawkins D (1988) Excavations at the Royal Mint Site 1986-1988. London Archaeologist 5: 429-436.

Hartz AJ, Fischer ME, Bril, G, Kelber S, Rupley D Jr, Oken B and Rimm AA. (1986) The association of obesity with joint pain and osteoarthritis in the Hanes data. Journal of Chronic Diseases 39: 311-319.

Harvey B (1993) Living and dying in England, 1100-1540: the monastic experience. Oxford: The Clarendon Press.

Julkunen H, Heinonen OP and Pyörälä K (1971) Hyperostosis of the spine in an adult population: its relation to hyperglycaemia and obesity. Annals of the Rheumatic Diseases 30: 605-612.

Larsen CS (1997) Bioarchaeology: interpreting behavior from the human skeleton. Cambridge: Cambridge University Press.

Leach RE, Baumgard S and Broom J (1973) Obesity: its relationship to osteoarthritis of the knee. Clinical Orthopaedics and Related Research 93: 271-273.

Lieberman LS (2000) Obesity. In KF Kiple and KC Ornelas (eds.): The Cambridge world history of food. Cambridge: Cambridge University Press: 1062-1077.

Lievense AM, Bierma-Zeinstra SMA, Verhagen AP, van Baar ME, Verhaar JAN and Koes BW (2002) Influence of obesity on the development of osteoarthritis of the hip: a systematic review. Rheumatology 41: 1155-1162.

McHenry HM (1975) Fossil hominid body weight and brain size. Nature 254: 686-688.

McHenry HM (1988) New estimates of body weight in early hominids and their significance to encephalization and megadontia in 'Robust' Australopithecines. In FE Grine (ed.): Evolutionary history of the 'Robust' Australopithecines. New York: Aldine de Gruyter: 133-148.

McHenry HM (1991) Sexual dimorphism in *Australopithecus afarensis*. Journal of Human Evolution 20: 21-32.

McHenry HM (1992) Body size and proportions in early hominids. American Journal of Physical Anthropology 87: 407-431.

Morse D, Duncan J and Stoutamire J (1983) Handbook of forensic archaeology and anthropology. Tallahassee: Morse, Duncan & Stoutamire.

Passmore R and Eastwood MA (1986) Davidson and Passmore human nutrition and dietetics. Edinburgh: Churchill Livingstone.

Patrick P (2005a) An archaeology of overindulgence. Archaeological Review from Cambridge 20.2: 98-117.

Patrick PJ (2005b) "Greed, gluttony and intemperance"? Testing the stereotype of the 'obese medieval monk'. Unpublished PhD thesis, University of London.

Porter AMW (1997) Physique and the skeleton. Unpublished PhD thesis: University of London.

Radin EL, Paul IL & Rose RM (1972) Role of mechanical factors in pathogenesis of primary osteoarthritis. The Lancet 1972: 519-522

Resnick D and Niwayama G (1976) Radiographic and pathologic features of spinal involvement in diffuse idiopathic skeletal hyperostosis (DISH). Radiology 119: 559-568.

Resnick D, Shaul SR and Robins JM (1975) Diffuse idiopathic skeletal hyperostosis (DISH): Forestier's disease with extraspinal manifestations. Radiology 115: 513-524.

Rogers J (2000) The palaeopathology of joint disease. In M Cox and S Mays (eds.): Human osteology in archaeology and forensic science. London: Greenwich Medical Media: 163-182.

Rogers J, Shepstone L and Dieppe P (1997) Bone formers: osteophyte and enthesophyte formation are positively associated. Annals of the Rheumatic Diseases 56: 85-90.

Rogers J and Waldron T (2001) DISH and the monastic way of life. International Journal of Osteoarchaeology 11: 357-365.

Rotés-Querol J (1996) Clinical manifestations of diffuse idiopathic skeletal hyperostosis (DISH). British Journal of Rheumatology 35: 1193-1194.

Ruff CB, Scott WW and Liu AYC (1991) Articular and diaphyseal remodelling of the proximal femur with changes in body mass in adults. American Journal of Physical Anthropology 86: 397-413.

Saville PD and Dickson J (1968) Age and weight in osteoarthritis of the hip. Arthritis and Rheumatism 11: 635:644.

Spector TD (1990) The fat on the joint: osteoarthritis and obesity. The Journal of Rheumatology 17: 283-284.

Steudel K (1980) New estimates of early hominid body size. American Journal of Physical Anthropology 52: 63-70.

Trotter M (1970) Estimation of stature from intact long limb bones in Stewart, T D (ed.) Personal identification in mass disasters. Washington: National Museum of Natural History, Smithsonian Institution; 71-83.

Ulijaszek SJ (1998) Obesity, fatness and modernization" in Ulijaszek, S J, Johnston, F E & Preece, M A (eds) The Cambridge encyclopedia of human growth and development. Cambridge: Cambridge University Press; 410-411.

Waldron T (1985) DISH at Merton Priory: evidence for a 'new' occupational disease? British Medical Journal 291: 1762-1763.

Waldron T (1987) The relative survival of the human skeleton: implications for palaeopathology. In A Boddington, AN Garland and RC Janaway (eds.): Death, decay and reconstruction: approaches to archaeology and forensic science. Manchester: Manchester University Press; 55-64.

Waldron T and Rogers J (1991) Inter-observer error in coding of osteoarthritis in human skeletal remains. International Journal of Osteoarchaeology 1: 49-56.

Wolff J (1892) The law of bone remodelling (trans. Maquet P & Furlong R). Berlin: Springer-Verlag.

A New Method for Recording Tooth Wear

Recording Tooth Wear

Anna Clement

Institute of Archaeology, UCL
31-34 Gordon Square
London
WC1H 0PY
a.clement@ucl.ac.uk

Abstract

Modern humans living in industrialised societies use their dentition for a limited number of functions. Consequently they often lack any visible evidence of tooth wear. The human dentition, however, has evolved to withstand heavy wear. Non-industrialised, modern and ancient populations exhibit evidence for much heavier patterns of tooth wear. This tooth wear provides us with information about many aspects of their lifestyle and behaviour, including diet, food preparation, age-at-death and craft activities. A wealth of literature has been published describing and testing various methods for recording tooth wear. This literature includes both ordinal and, more recently, quantitative scoring techniques. A new quantitative method is presented here. It combines the advantages of both quantitative and ordinal methods, while minimising their disadvantages. A pilot study was conducted on the skeletal remains from the medieval sites of the Royal Mint and Merton Priory from London, England. The dentitions of 30 adults from each site were measured and analysed. The results of this pilot study illustrate the accuracy of this new method and its potential applicability to large samples.

Keywords: Tooth Wear, Merton Priory, Royal Mint, Teeth.

Introduction and background

Today, humans who live in industrialised societies use their dentition for a very limited number of functions. The main cause of tooth wear in these modern groups is the abrasion of a tooth's enamel surface through repeated brushing, to keep them clean and free from disease (Hillson 1996). It, therefore, makes it hard to imagine a time when teeth had many different functions, both in the process of preparing food and in many craft activities.

Heavy tooth wear has been reported in both historic and prehistoric populations, as well as modern hunter-gatherers and has been the subject of detailed studies (Barret & Brown 1975; Bermúdez de Castro & Pérez 1986; Cook 1981; Hinton 1981; Mays 2002; Molnar et al. 1983). The morphology of the dentition has evolved to withstand heavy tooth wear and has developed mechanisms to protect the pulp cavity from exposure. The process of tooth wear is complex, but in simplified terms once the top layer of enamel is worn away, the dentine (a hard, yellow-coloured material, which often turns brown in a burial environment) is exposed. When this dentine is worn through, secondary dentine is produced to protect the pulp cavity. In cases of extreme wear this pulp cavity may become exposed, leading to the loss of the tooth through infection (for further detail see Hillson 1996).

There are three main types of tooth wear; erosion, attrition and abrasion. Erosion is the chemical dissolution of enamel and dentine by acids not produced by oral bacteria; attrition is the physical wearing of a tooth's surface, caused by the interaction between opposing and neighbouring teeth, resulting in the loss of enamel at locations of contact; and abrasion is the physical wearing of teeth resulting from mechanical grinding, rubbing, scraping or micro-cutting caused by the introduction of objects into the mouth (Bell et al. 1998). Teeth are rarely subjected to a single form of wear and normally cope with the accumulative damage from a diverse array of factors.

Patterns of tooth wear are influenced by a combination of biological, physical and cultural determinants, including: the position of individual teeth, the relationship between the upper and lower dental arches, the abrasiveness of food and any incorporated foreign material, the use of the dentition as tools, and intentional mutilation (Barret & Brown 1975). All of these factors can create unique wear patterns causing great differences between and within populations. Hinton (1981), for example, identified different tooth wear patterns within groups of Australian Aborigines, Inuit of Greenland and Canada, Native Americans and Ohio Woodlanders. These differences were attributed to the varying abrasiveness of their diets. Smith (1984) also noted that hard particles introduced into the diet through the grinding of cereals with stone tools during the Neolithic period caused an increase in the amount of wear and a change in the angle at which a tooth was worn.

The study of tooth wear patterns in both modern and ancient human populations has a long research history, the very earliest study being conducted by Broca at the end of the 19th century (Broca 1879). These early studies were motivated by the fact that, once recorded, tooth wear patterns can be used to make inferences about diet, food preparation techniques, and habitual activities involving the teeth. Teeth also form the most commonly preserved element of the human skeleton and also provide one of the most useful skeletal markers to age adult skeletons with (Hillson 1996). Consequently, a wealth of literature has been published describing and testing various methods for recording tooth wear (Brothwell 1981; Kambe et al. 1991; Miles 1962; Molnar et al. 1983; Richards 1984; Richards & Brown 1984; Scott 1979; Smith 1984; Walker et al. 1991). This literature includes both micro- and macro-wear, ordinal and quantitative techniques, and occlusal and non-occlusal studies.

Micro-wear studies involve the visual assessment and quantification of abrasive features, such as facets, pits and scratches, which appear on a tooth's surface, using a high-powered microscope, such as scanning electron microscope (SEM) (Teaford 1988; Teaford & Lytle 1996). Micro-wear features have a short temporal life, which provides an opportunity for assessing dietary change and even seasonality of death (Buikstra et al. 1994). While micro-wear analysis represents an exciting and relatively new development in tooth wear research, it does have several disadvantages. The short temporal life of micro-wear features can cause difficulties when trying to accurately reconstruct the diet of an individual if their food supply was highly seasonal. The features produced by hard foods also tend to predominate over less marked features produced by soft foods (Hillson 1996). The technique is extremely time consuming and is based on the enamel surfaces of teeth, so individuals with extreme wear, where very little enamel remains cannot be accurately examined using this method. These factors made it inappropriate for use on the two populations represented in the pilot study.

Macro-wear looks at the wear on the surface of a tooth that is visible to the naked eye. It is much less time consuming than micro-wear analysis and requires less specialised equipment. There are two main types of methods used to record occlusal macro-wear; ordinal and quantitative. Ordinal methods are based on a five-stage scale developed by Broca (1879). They involve the visual assessment of the amount of enamel and dentine present on each tooth. Others have developed variations of this method (e.g. Broca (1879), Molnar (1971), Scott (1979) and Smith (1984)), but the approach remains the same: the wear present on each tooth is placed into discrete categories, for example on a scale of 0-8. This type of technique has several advantages; it is easy to learn, speedily applied and requires no specialised equipment. However, the boundaries between each wear class are highly subjective, leading to the situation where two researchers could allocate a tooth showing the same

amount of wear to different categories (Walker et al. 1991). They also have statistical disadvantages, as ordinal data are discrete and non-continuous, preventing both the use of more powerful statistical tests and limiting the conclusions that can be drawn from the data (Walker et al. 1991).

Contrastingly, quantitative methods measure the area of occlusal wear facets from photographs of teeth using a planimeter or a computer-based image system (Kambe et al. 1991; Molnar et al. 1983; Richards 1984; Richards & Brown 1981; Walker 1978). They present a more accurate way of measuring tooth wear, minimising the amount of both inter- and intra-observer error. They also produce continuous data, allowing small differences between and within populations to be identified as well as a broader statistical use of the data. These methods have previously been criticised for being complex to learn, time consuming and often requiring specialised equipment.

Tooth wear can occur on all five of a tooth's surfaces, but is rarely recorded on any surface other than the occlusal (see above). This is primarily because non-occlusal tooth wear is highly variable and, therefore, difficult to standardise or categorise. Though non-occlusal tooth wear has been reported in detail, for features such as chipping and inter-proximal grooves (Berryman et al. 1979; Brown & Molnar 1990; Formicola 1988; Schulz 1977; Turner & Cadien 1969; Uberlaker et al. 1969), these reports are mostly descriptive. While many interesting, descriptive accounts of non-occlusal wear continue to be published there is, at present, no standardised method for recording the various types of non-occlusal wear. Until standard methods are developed comparisons of non-occlusal wear between and within populations remain problematic.

The new method for measuring tooth wear presented here is a 2-D macro and quantitative technique that records the wear on the occlusal surface of a tooth. It was developed to provide an accurate way of recording small differences in tooth wear patterns within and between different populations and combines the accuracy of previous quantitative methods with the ease of a recording system only previously known for ordinal methods. It also produces continuous data and minimises intra-observer error. This method was developed as part of a larger study on tooth wear patterns in Neanderthals and Early Modern Humans, but it is the results of the pilot study that are presented here. In the pilot study the new method was applied to a group of 60 medieval skeletons from the sites of Merton Priory and the Royal Mint in London, England.

The effects of age make it hard to observe variations in wear patterns at a population level. This is because tooth wear is strongly affected by the age of an individual, i.e. the older the individual, the heavier wear. A technique

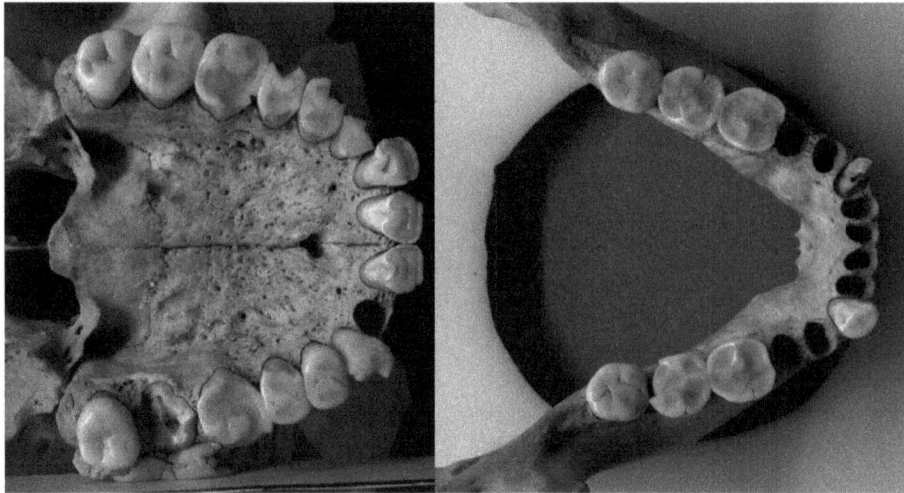

Figure 1. Digital photographs of the occlusal surfaces of the maxillary and mandibular dentition.

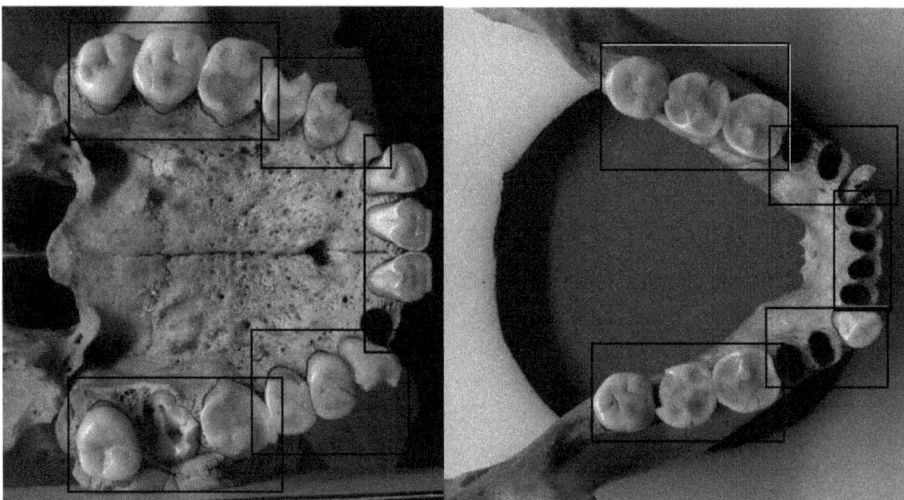

Figure 2. Digital photographs of the occlusal surfaces the maxillary and mandibular dentition. The rectangles denote the sections that are isolated for measurement.

was therefore developed during the pilot study to remove most of the effects of age at both an individual and population level. This new method and technique are both discussed below.

Method

This section describes the new method developed for recording tooth wear, and explains how results are stored/processed. Firstly, digital photographs are taken of the occlusal surface of the mandibular and maxillary teeth of an individual. An example of this is shown below (Fig. 1). Each photograph of the mandible or maxilla is divided into five sections for measurement (Fig. 2). These five sections comprise; the left first molar (M1), second molar (M2) and third molar (M3); the left fourth premolar (P4), third premolar (P3) and canine (C); the left and right incisors (I1s and I2s); the right P4, P3 and C; and the right M1, M2 and M3. Each section is isolated from the main photograph and enlarged to the

size of 20cm along its longest edge. Measurements are then taken from the teeth within each section.

Each tooth is measured using the computer software program, Sigma Scan Pro. A graphics tablet is used to draw around the occlusal outline of each tooth and calculate its area in pixels. The darker area of dentine is then measured using the same method. Any isolated patches of dentine on the occlusal surface of a tooth are measured individually and then added together to calculate the total area. To calculate the ratio of wear the total area of dentine is divided by the area of the occlusal surface. Pixels are used instead of metrics to calculate the area of wear, which dispenses with the need for a scale in each picture. Metrics are not needed because it is the ratio of areas that is being recorded and not their actual size.

The area of the occlusal surface (representing the original area of the enamel) and dentine as well as the wear ratios for each individual are then recorded in separate Excel

Table 1. An example of the excel spreadsheet used to record the area of dentine and enamel (occlusal area) in the teeth from the left side of the maxilla and mandible for one individual. The ratio is also calculated at the bottom of the table for both the maxilla and the mandible. X represents either a damaged of missing tooth.

Maxilla	LM3	LM2	LM1	LP4	LP3	LC	LI2	LI1
Enamel	X	29533	38015	15331	12551	7314	2048	3824
Dentine	X	415	6944	736	201	937	193	1132

Mandible	LM3	LM2	LM1	LP4	LP3	LC	LI2	LI1
Enamel	15017	22721	29419	PM	PM	8720	4125	3642
Dentine	0	434	9002	PM	PM	1009	747	1053

Ratio	LM3	LM2	LM1	LP4	LP3	LC	LI2	LI1
Maxilla	X	0.0141	0.1827	0.0480	0.0160	0.1281	0.0942	0.2960
Mandible	0	0.0191	0.3060	X	X	0.1157	0.1811	0.2891

Table 2. An example of the excel spreadsheet used to group the ratios for the left maxillary teeth from a group of individuals. X represents either a damaged or missing tooth.

Individual	ULM3	ULM2	ULM1	ULP4	ULP4	ULC	ULI2	ULI1
6225	X	X	0.0765	0	0.0472	0.2350	0.2100	0.3279
6483	X	0.0141	0.1827	0.0480	0.0160	0.1281	0.0942	0.2960
7032	0	0.0165	0.0535	0	0.0081	0.2552	0.2950	X
7294	X	X	0.0852	X	0.0107	0.2306	0.0826	0.1780
7358	X	0.1423	0.1464	0.0484	0.0660	X	0.2904	0.2927
7425	X	0.0192	0.0945	0.0362	0.0663	0.3200	0.3378	0.4383
8012	X	X	0.3001	0.0249	0.0115	0.0784	0.1222	0.1614
8102	0	0.0207	0.0816	0.0051	0	X	X	X
8112	0.0086	X	0.3608	0.0922	0.0284	0.1090	0.1586	0.3585

spreadsheet (Table 1). The wear ratios calculated for each specimen within a given population are then grouped together in a separate excel spreadsheet (Table 2). These results are then copied into an SPSS data sheet for further analysis.

Left and Right

It is common in tooth wear studies to only measure the left or the right side of the dentition (Lukacs & Pastor 1988; Miles 1962; Molnar et al. 1983; Richards & Brown 1981; Smith 1984), because wear is often symmetrical. Archaeological specimens, however, often do not possess a complete dentition. During data collection for the pilot study, individuals were selected at random, but those who possessed less than half of their full dentition were not included in the study. Despite this selection criteria many of the individuals included in the study had lost teeth both ante- and post-mortem. Additionally, some of the teeth exhibited signs of dental disease, such as caries and dental calculus. If these conditions affected the occlusal surface, the tooth was excluded from the study.

It was therefore an advantage to be able to substitute an antimere (an opposite part) for a missing or unmeasurable, but it was first necessary to test for symmetry of wear.

A Mann-Whitney test was conducted on measurements taken on teeth from the left and right sides of the upper and lower jaws of the 60 individuals included in the pilot study. No significant differences were found between any of the teeth from the left and right sides of the upper and lower jaws at the usually accepted 5% level of significance (in fact the values are all much higher than this). It was therefore decided that wear ratios for the left and right teeth could be combined. Where both the left and right teeth are present the average of the two scores is taken. Where only one of the pair of teeth is actually present, the wear ratio is recorded for the tooth present.

Intra-observer error

An intra-observer error test was also applied to the 60 individuals included in the pilot study. The areas of enamel and dentine were calculated for the teeth within maxilla and mandible, using the method described above, for each individual. The area of the occlusal surface (represented the original enamel) and dentine for each tooth was measured three times over intervals of one week and a wear ratio was calculated each time. The measurement error was then calculated from these results and found to be 1.9%, substantially below acceptable 3% level.

The Royal Mint and Merton Priory

This new method was applied to 30 individuals, selected at random, from each of the medieval sites of Merton Priory and the Royal Mint that are both located in London, England. Both of these sites have a monastic context, which explains the predominance of male skeletons within both samples (Table 3).

Table 3. Sex distribution of the individuals selected from Merton Priory and the Royal Mint collections (WORD).

Sex	Merton Priory	Royal Mint
1 (Male)	19	18
2 (Male?)	8	4
3 (Indeter.)	2	2
4 (Female?)	1	4
5 (Female)	0	2

Table 4. Age distribution of the individuals selected from Merton Priory and the Royal Mint collections (WORD).

Age	Merton Priory	Royal Mint
7 (18-25)	0	3
8 (26-35)	7	9
9 (36-45)	14	15
10 (46 +)	7	3
11 (indeter.)	2	0

Merton Priory

During the excavation of this site, which was largely undertaken by a team from the Museum of London, a total of 738 medieval burials were discovered (Miller et al. in prep.). The burials were all associated with Merton Priory which was an Augustinian Priory founded in the early 12th century. It lasted until its dissolution in 1538, which led to the immediate demolition of the monastic buildings (Bruce & Mason 1993). The burials were discovered within four main areas: the lay cemetery, the church, the chapter house and the canons' cemetery. These burials were divided into four main phases dating between AD 1117 and 1538 (Miller et al. in prep.).

Basic information, including the age and sex of 30 adults from this site, was extracted from the Wellcome Osteological Research Database (WORD) (Tables 3 and 4). The sample consists of 27 males, 1 female and 2 sex indeterminates, who range in age from 26 to 46+.

The Royal Mint

Excavations at the site of the Royal Mint uncovered parts of two cemetery sites: a large catastrophe cemetery, associated with the Black Death; and a monastic cemetery, associated with the Cistercian Abbey of St Mary Graces. Seven-hundred-and-fifty burials were recovered from the catastrophic cemetery, and 420 from the monastic cemetery (Grainger et al. 1988). The association of the catastrophe cemetery with the Black Death gives it a precise date of AD 1348 to 1350 (Grainger et al. 1988). The monastic cemetery is dated from AD 1350 to 1540 (Grainger et al. 1988).

Basic information about the age and sex of 30 adults from this site was, again, extracted from the Wellcome Osteological Research Database (WORD) (Tables 3 and 4). This sample consists of 22 males, 6 females and 2 sex indeterminates, who range in age from 18 to 46+ years.

Results and Discussion

Tooth wear is usually recorded as an overall amount of wear (Buikstra et al. 1994; Hinton 1981; Kieser et al. 1985; Lovejoy 1985; Smith 1984). This type of information is particularly useful in determining the age of individuals within a population, but does little to provide valuable information about wear patterns. This is because tooth wear is highly correlated with age, i.e. the older an individual the higher the amount of wear. Age, therefore, often obscures important wear patterns and needs to be removed as a factor.

Detailed studies have been made on the eruption sequence of adult teeth in modern humans (Jaswal 1983; Schour & Massler 1941; Smith & Garn 1987; Ubelaker 1978). From these studies a normal eruption sequence has been established for the adult human dentition:

Upper dentition: M1,I1,I2,P1,C,P2,M2,M3
Lower dentition: M1,I1,I2,C,P1,P2,M2,M3

Their emergence times are grouped into three phases (Smith & Garn 1987):

Phase 1: Emergence of permanent first molars and incisors (5-8 years).
Phase 2: Emergence of canines, premolars and second molars (9.5-12.5 years).
Phase 3: Emergence of third molars (late teens/ early twenties).

These eruption sequences are important because they determine how long each tooth would have been exposed to wear. If a constant rate of wear is assumed for the whole dentition, then the amount of wear exhibited by each tooth should closely match its position within the eruption sequence; the first molar being the most worn and the third molar being the least.

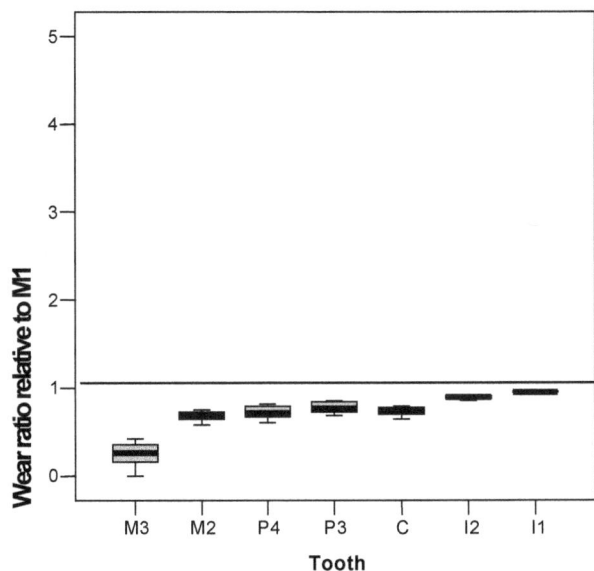

Figure 3. This graph present the expected wear ratios for each tooth, divided by the expected wear ratio of the M1 (the wear ratio relative to the M1). The horizontal line passing through 1 on the y-axis represents the M1 wear ratio.

Tooth wear could be calibrated against the age of the individual, but in practice age is unknown in most archaeological populations and precise age estimations using the skeletons are often difficult (frequently relying on tooth wear patterns), so it is better to use an alternative approach to remove age as a factor. This can be done be standardising the wear for each tooth, relative to the wear of one particular tooth in the dentition. The wear ratio of each tooth is presented relative to that of the M1. The M1 was chosen because it is the first tooth in the permanent dentition to erupt and come into occlusion.

If a constant rate of wear is assumed then an expected wear pattern can be calculated from the average eruption timings of teeth. Figure 3 illustrates this expected wear pattern, with an assumed wear rate of 1% per year. The wear ratios for each have been divided by the wear ratio of the M1, so the wear ratio of each tooth is now presented relative to that of the M1. Each box plots represent an individual tooth, starting with the back of the mouth (M3) following through to the front (I1). The ends of each box are the upper and lower quartile values, and so the height of the box represents the inter-quartile range. The line inside the box represents the median, and the 'whiskers' represent the overall range of values.

Outliers are represented by 'O' and 'X' symbols. The horizontal line, passing through 1 on the y-axis, represents a state of wear equivalent to wear on the M1.

When this expected wear pattern (Fig. 3) is compared to the results for the sample populations from Merton priory and the Royal Mint (Figs. 4 and 5) inter-population differences and similarities in wear patterns begin to emerge. Both sample populations possess a number of outliers. These outliers represent wear ratio values that are far removed from the other wear ratio values for that tooth within the same sample. Outliers can be the result of measurement error, but in this study each outlier was checked and all represent true values. The outliers for the Royal Mint group represent a mixture of adjacent teeth in particular individuals, such as skeleton 13 and skeleton 27, and isolated teeth from different individuals. There is also a higher number of outliers in the lower dentition compared to the upper dentition, indicating that most of the upper teeth fit within a more standardised wear pattern in the Royal Mint sample sample. Contrastingly, most of the outliers in the Merton Priory sample represent specific individuals rather than individual/adjacent teeth. Individual 44, for example, has outliers for their upper M3s and lower M3s, M2s and Cs.

These outliers represent teeth/individuals that possess wear ratios that are much higher than those of the majority of ratios within the same sample, i.e. these teeth are much more heavily worn, relative to the M1, than would be expected from their position in the eruption sequence. The higher number of outliers in the Merton Priory sample suggests that a greater variation in wear ratios exists within this group, compared to those from the Royal Mint site.

The upper dentitions

While the upper teeth, in both the Merton Priory and Royal Mint samples, possess similar wear patterns to those expected from their position within the eruption sequence, subtle differences are also visible.

All of the teeth from the Royal Mint sample, except the first incisors, have medians below the M1 wear ratio line (Fig. 4). The posterior teeth have low wear medians, located just above the 0 level (Fig. 4). This indicates that the posterior teeth possess very small amounts of wear, relative to the M1. In contrast, the anterior teeth have a

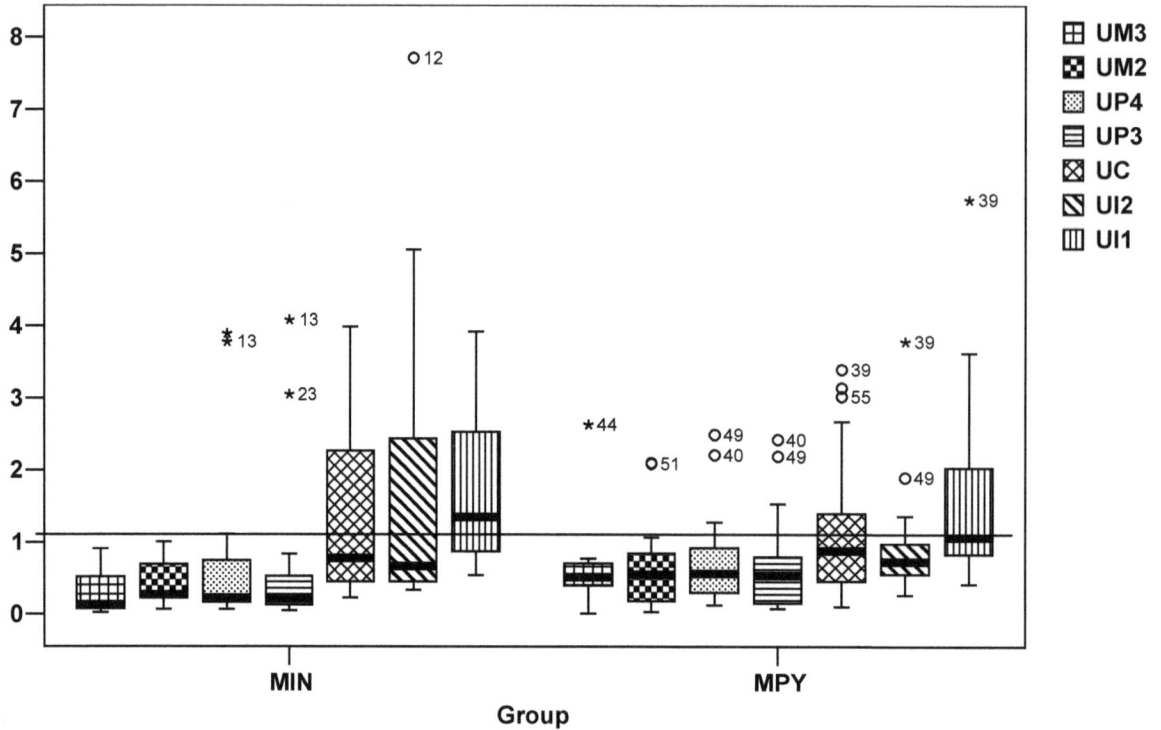

Figure 4. Wear ratios for the upper teeth of the Royal Mint (MIN) and Merton Priory (MPY) groups, divided by the wear ratio of the M1. The horizontal line passing through 1 on the y-axis represents the M1 wear ratio.

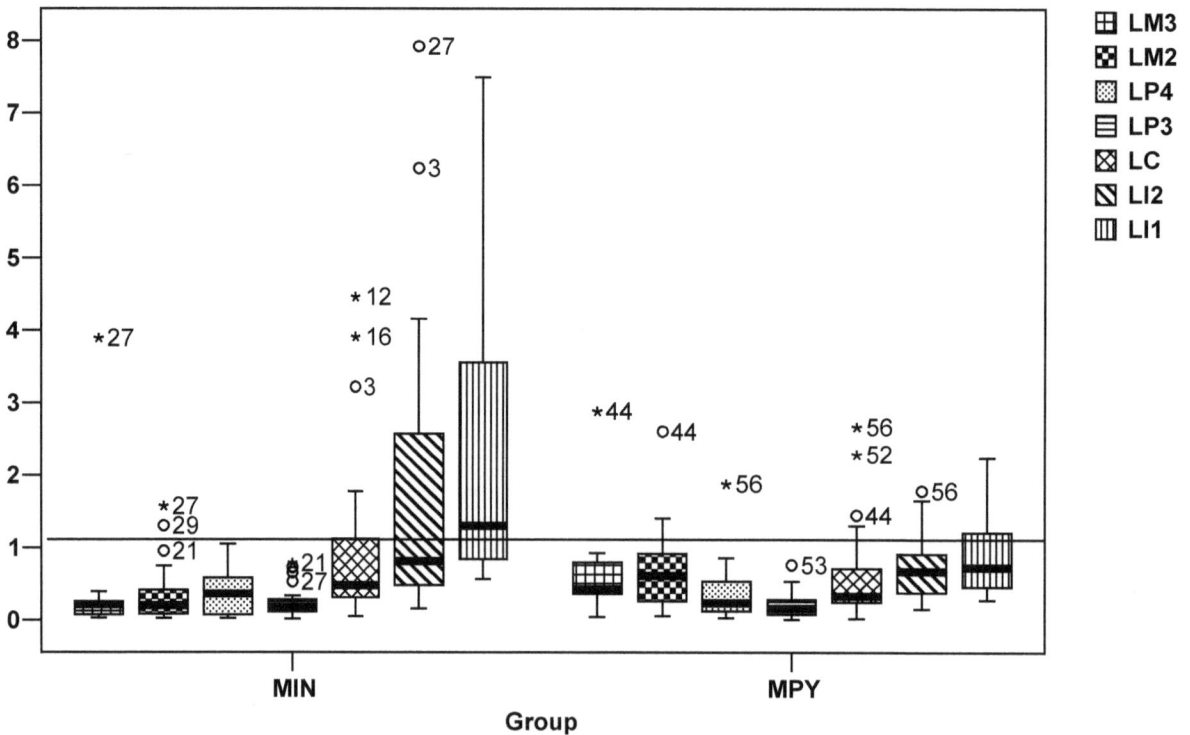

Figure 5. Wear ratios for the lower teeth of the Royal Mint (MIN) and Merton Priory (MPY) groups, divided by the wear ratio of the M1. The horizontal line passing through 1 on the y-axis represents the M1 wear ratio.

much higher and wider range of wear ratios. The upper limits of their inter-quartile ranges are situated above the level representing twice the amount of wear of the M1 (Fig. 4). The upper limits of the whiskers reach, or extend beyond, the quadruple level of the M1 wear ratio (Fig. 4). This indicates that many of the individuals within this sample possess heavier amounts of anterior tooth wear than expected from their position within the dental eruption sequence.

The upper teeth from Merton Priory all have medians on or below the M1 wear ratio line (Fig. 4). They also possess small ranges in their wear ratios, relative to the M1. This represents a more universal wear pattern shared amongst the majority of adults within this group. The most prominent feature of this wear pattern is the relatively large amounts of wear on the posterior teeth. These teeth possess heavier amounts of wear, relative to the M1, than the Royal Mint group (Fig. 4). This is especially interesting for the M3, which has a wear median similar to that of the P3, P4 and M2. This is surprising as the M3 erupts significantly later than the other posterior teeth. It strongly suggests that the adults from Merton priory were exposing their posterior dentition to higher amounts of wear than the adults from the Royal Mint site.

The lower dentitions

As expected, the lower teeth from the Royal Mint and Merton Priory groups show similar results to their upper counterparts (Figs. 4 and 5). They possess patterns of wear that broadly reflect their individual positions within the eruption sequence.

All of the lower teeth from the Royal Mint group, except for the I1s, have medians below the M1 wear ratio line (Fig. 5). The posterior teeth have exceptionally low wear ratios, relative to the M1. The lower anterior teeth mirror the wear pattern seen in the upper teeth, with inter-quartile ranges rising above the M1 wear line (Fig. 5). The upper inter-quartile ranges of the I1s and I2s extend above the level which represents twice the amount of wear of the M1. The upper whisker of the I1 reaches as high as seven-times the wear ratio of the M1 (Fig. 5). These results, combined with those from the upper teeth, strongly support the idea that many of the individuals within this group were wearing their teeth at a faster rate than that expected from the eruption sequence. This further supports the suggestion that these anterior teeth were being use for a non-masticatory function, possibly cultural activities.

The lower teeth from the Merton Priory group all possess medians that fall below the M1 wear ratio line (Fig. 5). They possess even smaller ranges in their wear ratio values than their upper counterparts. The posterior teeth show higher amounts of wear than predicted from their positions within the eruption sequence. This is most noticeable in the M2s and M3s (Fig. 5). This is also a feature of the upper teeth, and suggests that the individuals from Merton Priory may have been using their posterior teeth, either to masticate a more abrasive diet than those from the Royal Mint site, or in craft activities.

Conclusions

This paper demonstrates how a new method for recording tooth wear can be used to; record the amount of wear in individuals, or groups of individuals; compare the wear patterns between selected groups: and has the potential to help define and answer specific research questions (for example, on the cultural use of teeth). This is largely due to the accuracy of this quantitative method and because it produces continuous data. More complex statistical analysis of this data is beyond the scope of this paper, but will be applied during further and more in-depth analyses of these results.

While the wear patterns extracted from Merton Priory and the Royal Mint collections largely resemble those predicted from the adult eruption sequence, the accuracy of this new method and the removal of age as a major factor, allows more subtle differences between the two populations to be identified. The most striking features of their tooth wear patterns are the high range and level of wear ratios in the anterior dentition of the Royal Mint group, and the heavy posterior wear in the Merton Priory group. Currently we can only postulate that either some form of cultural use of these teeth, or their use in the mastication of abrasive foodstuffs, led to these differences. Future research will investigate these differences in more detail and assess their possible causes. This will primarily involve extending the sample size of each group, and a detailed analysis of the associated archaeology and historical documents. It is also the current intention to calculate the inter-observer error for this new method.

This technique, unlike previous quantitative methods, does not require specialist equipment and can be more speedily applied, making it suitable for use on larger collections. If time restrictions are encountered when recording larger collections of human skeletal material, digital photographs could be taken and stored for future analysis.

This new method for recording and calibrating tooth wear has the potential to make a substantial contribution to the way in which we record and analyse tooth wear in archaeological collections, ranging from fossil hominids to modern groups.

Acknowledgements

I am grateful to Bill White and Tania Kausmally from the Centre of Human Bioarchaeology at the Museum of London for their help with, and access to, the Merton Priory and Royal Mint collections. I would also like to thank Professor Simon Hillson and Dr Charles Fitzgerald for their assistance in developing this new method for recording tooth wear. This research was supported by the Arts & Humanities Research Council (Grant #101875).

Literature Cited

Barret JA and Brown T (1975) Comments in Wallace JA: Did La Ferrassie I Use His Teeth as a Tool? Current Anthropology 16: 393-441.

Bell EJ, Kaidonis JA, Townsend GC and Richards L (1998) Comparison of exposed dentinal surfaces resulting from abrasion and erosion. Australian Dental Journal 43: 362-366.

Bermúdez de Castro JM and Pérez PJ (1986) Anomalous tooth-neck wear in North African Mesolithic populations. Paleopathology Newsletter 54: 5-10.

Berryman HE, Owsley DW and Hendersen AM (1979) Non-carious interproximal grooves in Arikara Indian dentitions. American Journal of Physical Anthropology 50: 209-212.

Broca P (1879) Instructions relative á l'étude anthropologique du système dentaire. Bulletin Society Anthropology 36: 381-390.

Brothwell DR (1981) Digging up Bones. British Museum: London.

Brown T and Molnar S (1990) Interproximal grooving and task activity in Australia. American Journal of Physical Anthropology 81: 545-553.

Bruce P and Mason S (1993) Merton Priory. MoLAS: Mitcham.

Buikstra JE, Frakenberg S, Lambert JB and Xue LA (1994) Standards for Data Collection from Human Skeletal Remains. Arkansas Archaeological Survey: Fayetteville.

Cook DC (1981) Konig Eskimo Tooth Ablation: Was Hrdlicka Right After All? Current Anthropology 22: 159-163.

Formicola V (1988) Interproximal Grooving of Teeth: Additional Evidence and Interpretation. Current Anthropology 29:663-671.

Grainger I, Hawkins D, Falcini P and Mills P (1988) Excavations at the Royal Mint Site 1986-1988. London Archaeologist 5: 429-436.

Hillson S (1996) Dental Anthropology. Cambridge University Press: UK.

Hinton RJ (1981) Form and patterning of anterior tooth wear among Aboriginal human groups. American Journal of Physical Anthropology 67: 393-402.

Jaswal S (1983) Age and sequence of permanent tooth-emergence among Khasis. American Journal of Physical Anthropology 62: 177-186.

Kambe T, Yonemitsu K, Kibayashi K and Tsunenari S (1991) Application of a computer assisted image analyzer to the assessment of area and number of sites of dental attrition and its use for age estimation. Forensic Science International 50: 97-109.

Kieser JA, Groeneveld HT and Preston CB (1985) Patterns of dental wear in the Lengua Indians of Paraguay. American Journal of Physical Anthropology 66: 21-29.

Lovejoy CO (1985) Dental wear in the Libben population: its functional pattern and role in the determination of adult skeletal age at death. American Journal of Physical Anthropology 68: 47-56.

Lukacs JR and Pastor RF (1988) Activity-induced patterns of dental abrasion in prehistoric Pakistan: evidence from Mehrgarh and Harappa. American Journal of Physical Anthropology 76: 377-398.

Mays S (2002) The relationship between molar wear and age in an early 19th century AD archaeological human skeletal series of documented age at death. Journal of Archaeological Science 29: 861-871.

Miller P, Saxby D and Coheeny J (In prep.) Excavations at the Priory of St. Mary Merton, Surrey. MoLAS monograph.

Miles AEW (1962) Assessment of the ages of a population of Anglo-Saxons from their dentitions. Proceedings of the Royal Society of Medicine 55: 881-886.

Molnar S (1971) Human tooth wear, tooth function and cultural variability. American Journal of Physical Anthropology 34: 175-190.

Molnar S, McKee JK and Molnar I (1983) Measurements of tooth wear among Australian Aborigines: serial loss of the enamel crowns. American Journal of Physical Anthropology 61: 51-65.

Richards LC (1984) Principal axis analysis of dental attrition data from two Australian Aboriginal populations. American Journal of Physical Anthropology 49: 271-276.

Richards LC and Brown T (1981) Dental attrition and age relationships in Australian Aboriginals. Archaeology Oceania 16: 94-98.

Schour I and Massler M (1941) Studies in tooth development: the growth pattern of human teeth. Journal of the American Dental Association 27: 1778-1792.

Schulz PD (1977) Task activity and anterior tooth grooving in prehistoric Californian Indians. American Journal of Physical Anthropology 46: 87-92.

Scott EC (1979) Principal axis analysis of dental attrition data. American Journal of Physical Anthropology 51: 203-212.

Smith BH and Garn SM (1987) Polymorphisms in eruption sequence of permanent teeth in American children. American Journal of Physical Anthropology 74: 289-303.

Smith HS (1984) Patterns of molar wear in hunter-gatherers and agriculturalists. American Journal of Physical Anthropology 63: 39-56.

Teaford MF (1988) A review of dental micro-wear and diet in modern mammals. Scanning Microscopy 2: 1149-1166.

Teaford MF and Lytle JD (1996) Diet-induced changes in rates of human tooth microwear: a case study involving stone-ground human maize. American Journal of Physical Anthropology 100: 143-147.

Turner CG and Cadien JD (1969) Dental chipping in Aleuts, Eskimos and Indians. American Journal of Physical Anthropology 31: 303-310.

Ubelaker DH (1978) Human Skeletal Remains: Excavation, Analysis and Interpretation. Aldine: Chicago.

Walker A, Dean G and Shapiro P (1991) Estimating age from tooth wear in archaeological populations. In MA Kelley and CS Larsen (eds.): Advances in Dental Anthropology. New York: Wiley-Liss; 169-178.

Walker PL (1978) A Quantitative analysis of dental attrition rates in the Santa Barbara Channel area. American Journal of Physical Anthropology 48: 101-106.

Intra-Community Variation in Uptake of a New Staple Crop in the Eastern Woodlands of North America

Fionnuala Rose

Rutgers, the State University of New Jersey, USA

Current address: 12 Oaklands
83 Penhill Road
Lancing
West Sussex
BN15 8HB
Email: f.rose@uclmail.net

Abstract

This study investigated uptake of an introduced staple crop, maize, into an indigenous centre of horticultural development at temperate latitudes. Carbon isotope ratios of human bone collagen were analysed from five sites in North America, from the central Mississippi and lower Illinois River valleys in Illinois, and contrasted with the wider Eastern Woodlands culture area. Adoption of the new crop has been linked to ritual or status-related activities, but despite clear status differences in Middle Woodland-era mortuary practices, no dietary differences were identified between individuals. Evidence for substantial consumption of maize was not expected till after AD 800, but was identified considerably earlier, repeatedly noted in the early Late Woodland, possibly as early as AD 400. As maize utilization increases, a marked and persistent bi-modal pattern is apparent in consumption of old and new crops at all sites. Maize consumption was not correlated with gender, biological age, or relative age of burials dated with fluoride, and the bi-modal pattern persisted for several centuries, indicating that there was true intra-community variation in uptake of the new crop, over a long period. This bi-modal pattern may be explained by differing properties of maize and indigenous seed crops, and their suitability for differing household requirements related to relative fertility, population density and social role.

Keywords: prehistoric diet, North America, Eastern Woodlands, Illinois, carbon isotopes, fluoride dating, maize, corn

Introduction

Diet and food-getting are recognised as fundamental aspects of prehistoric life, and dietary transitions have long been of interest to archaeologists. On a wider scale, increased dependence on crop growing has often broadly coincided with fundamental changes in occupation practice and in social organization. There are many disadvantages associated with a focus on horticulture, however, and it is not clear whether new practices were adopted through opportunity or necessity. Some recent studies have attempted to resolve these questions by breaking the community down into a collection of individuals with different needs and agendas, and have highlighted issues related to intra-community variation, differential access, and which members of society led the way in adopting new practices (Watson and Kennedy 1991; Hastorf 1998).

The Eastern Woodlands region of North America is unusual in that it experienced both indigenous (primary) and introduced (secondary) horticultural regimes. Crop plants of the indigenous Eastern Agricultural Complex started to evolve in the late Archaic as early as 4500 years ago, and continued to dominate even after a new crop, maize, was introduced via the Southwest c.2000 years ago. Although maize was to become dominant in most areas of the Eastern Woodlands, it was not until the late Late Woodland period post-AD 800 that there is widespread evidence for its more intensive exploitation.

This study focuses on the transition to maize growing within communities already practising horticulture, looking at differences in uptake within the community at initial adoption, and during its subsequent increase in popularity, to shed light on the process of crop adoption. It is believed that communities switched to the new crop over a short period, but there are few studies that have studied the period of transition itself. In order to investigate this, multiple temporal components were analysed from five different sites within a small geographical region

Some of the earliest substantial botanical finds of maize in the Eastern Woodlands come from west-central Illinois in the American Midwest. Most of the sites in this study are in the central Mississippi River valley in west-central Illinois (CMVI), to the north of the American Bottom region, location of the network of sites culminating in the great pyramidal centre Cahokia. The 'backwater' CMVI is an area from which little bioarchaeological data has previously been published. In addition to providing new information, the CMVI data can be compared with the adjacent lower Illinois River valley (LIV), an area with a well-established bioarchaeological research tradition, with extensive skeletal series and well-described archaeology (e.g. Cook 1980, Buikstra 1984). The study area therefore presents an excellent opportunity to investigate the timing and pattern of the transition to what would become a new staple crop.

Isotopic analysis of human remains was used to study changes in diet. Investigation with isotopes provides unique information. Most evidence for prehistoric diets is qualitative, using floral and faunal remains, however these are inconsistently preserved, complicating quantification of particular components of diet. Neither does study of such remains usually allow the archaeologist to distinguish the diets of individuals, or to comment on intra-community variation. These goals can be achieved by analysis of light stable isotopes of carbon in human bone collagen, which can identify the appearance of maize in the diet.

Nitrogen isotope ratios were also analysed in this study, however since there are few reported results with which to compare them, and as their significance has mostly to do with local ecology, they will be reported elsewhere (Rose in prep.).

The Eastern Woodlands of North America

The Eastern Woodlands culture area stretched from the edge of the Great Plains in the west, across temperate North America to the east coast. Little skeletal material is available from the Archaic or Early Woodland in the CMVI, and none was included in this study.

It was long believed that plant cultivation was absent from the Eastern Woodlands until the diffusion into hunting-and-gathering societies of the 'classic' triumvirate of maize, beans, and squash from Mesoamerica (Smith 1987). In the last 20 years, however, it has become clear that the Eastern Woodlands were an area of indigenous agricultural development, with incipient horticulture and even domestication of native plants dating as far back as the Late Archaic hunting-and-gathering period (2500-1000BC) (Yarnell 1993). The suite of indigenous crop plants is known as the Eastern Agricultural Complex (Yarnell 1994), mostly annuals, from which the seeds were exploited. The earliest known is the squash, *Curcurbita pepo*, with the subsequent addition of carbohydrate-rich marsh elder/sumpweed (*Iva annua)*, goosefoot (*Chenopodium spp.*), and maygrass (*Phalaris caroliniana)*, and oily sunflower (*Helianthus annuus)* (Asch and Asch 1985; Yarnell 1993).

Maize (*Zea mays*) first appears in the archaeobotanical record in the Eastern Woodlands around 2000 years ago, probably via the Southwest, however is only found intermittently, and in very small quantities, for over a thousand years (Chapman and Crites 1987; Conard, et al. 1984; Riley, et al. 1994). In view of its subsequent economic importance, it is not clear why there was such a long time lag before maize use increased. Indigenous seed crops can produce high yields, have a better nutritional profile than maize, and were less labour intensive to raise (Asch & Asch 1985; Gremillion 2004; Simms 1987; Gallagher 1989, 1992; Smith & Cowan 2003) which may explain the delay. Early maize-growing experiments may have been unsuccessful, and the crop may have been introduced more than once (Hart 1999).

By the Middle Woodland or 'Hopewell' period (100 BC-AD 250) there is considerable evidence for hierarchical societies, with extensive trade networks, exotic materials, and distinctive artefact styles. Although there is controversy about how segmented Middle Woodland society was, and about how rank was acquired (Braun 1979; O'Brien 1986), grave goods and tomb location within burial mounds give clear evidence for ranking among individuals, with high status individuals buried in log-lined crypts, whilst others were secondary bundle burials in the surrounding mound (Brown 1979; Charles 1995; Tainter 1977).

This intra-community variation may have extended to access to certain types of foods. In the study area, high status has been found to correlate with higher protein intake, using nitrogen isotope ratios at Klunk (Schober 1998), and trace elements (a less reliable method) at Gibson (Lambert, et al. 1979), and also with higher achieved height in males at both sites (Buikstra 1976). Despite the use of Eastern Agricultural Complex plant foods, hunting, fishing, and nut collecting continued to play an important role in diet (Asch and Asch 1985; O'Brien and Pulliam 1996). Only occasional, isolated finds of maize have come from this period (Conard, et al. 1984). Botanical evidence for maize has repeatedly come from unusual contexts, in association with non-food items such as tobacco – for example at Mund in the American Bottom *c.* AD 500 (Hastorf and Johannessen 1994) – which may be a reflection of ritual rather than subsistence use, possibly including restriction to certain sections of society. There was a similar lag in maize consumption in South and Central America, with the crop apparently used initially in a ritual drink (*chicha*) (Johannessen 1993; Wymer 1994).

The Late Woodland period was a time of rapid change and considerable variability. The fine artefacts and long distance trade of the Middle Woodland disappear, and burials exhibit few status distinctions, with little in the way of grave goods. The period is traditionally divided into early and late Late Woodland. Early Late Woodland (AD 250-700) sites are smaller, with occupation generally more concentrated in major river valleys (Charles 1995; Connor 1991; O'Brien 1987). Subsistence continued to focus on collection and/or cultivation of native seed plants, nuts (Asch and Asch 1985; Johannessen 1993), and hunting and fishing, although resource acquisition may have been more localized at this time (Styles 1981; O'Brien 1987). Against this background, the Missionary Island No. 4 site, to the east in Ohio, stands out with carbon isotope ratios indicating regular maize consumption surprisingly early on, not long after AD 500 (Stothers and Bechtel 1987). In our study area, there are repeated botanical finds of small quantities of maize in the early Late Woodland at Deer Track, Buffalo, and Scenic Vista in the Sny Bottom area of the CMVI, and at Axedental and Elledge in the uplands between the CMVI and the LIV (Simon 2000). Carbon isotope ratios from LIV Koster Mounds (AD 600-800) also indicated consistent maize consumption (Schober 1998). These occurrences are notable not only for their early date, but also because these areas were not the main population centres, but were somewhat peripheral.

By the late Late Woodland (AD 700-1100) the number and size of habitation sites increased, and there was a renewed occupation of upland and secondary riverine sites (Charles 1995). Grave goods are still minimal, and there is again little evidence for status differences. Archaeobotanical evidence indicates that horticulture was now providing a major portion of food intake (Asch and Asch 1985; Johannessen 1993; O'Brien 1987). It is during this period that archaeobotanical maize remains are first recovered in any quantity across the Eastern Woodlands, and previous investigation with carbon isotopes has confirmed its increasing importance in diet at many sites post-AD 800 (van der Merwe and Vogel 1978; Buikstra 1992, Buikstra, et al. 1987; Schober 1998).

It is not clear why maize now rose in popularity and economic importance. It does have certain practical advantages over the small-seeded native cultigens, the single large seed head being easier to harvest and process. These properties may have reduced opportunity costs, particularly during the critical Fall period when other important resources - deer and nuts - were available (Gremillion 1996; Smith 1987). There is evidence for population growth at this time, which may have resulted in a need for increased crop productivity. Indigenous crop horticulture may itself have raised 'carrying capacity' and set in motion a population increase in some areas. Development of the Eastern eight-row/Northern Flint varieties did increase maize productivity, but Eastern eight-row is mainly found to the east in the Fort Ancient and Northeast areas, while older twelve-row varieties were the norm in this region (Fritz 1990; Wagner 1994). Finally, maize can be eaten many different ways, in its green state is sweeter and tastier than the rather bland indigenous crops (Smith and Cowan 2003).

The classic Mississippian (800-1650 AD) is characterized by a hierarchical settlement pattern culminating in large ceremonial centres with high population densities. The best-known example, Cahokia with its large pyramidal mounds, is in the American Bottom nearby to the south of the study region. Mississippian peoples were supported by extensive use of domesticated plants, notably maize, although it is not clear what, if any, role maize played in the socio-cultural changes associated with Mississippian (Wymer 1993). In the CMVI and LIV, however, the nature of 'Mississippian' occupation is not well understood: at this northernmost edge of the Mississippian culture area, 'Mississippian' sites are comparatively few, and late Late Woodland occupation appears not only to have overlapped in time and space, but to characterize the majority of archaeological remains from in this region (Farnsworth, et al. 1991; Muller and Stephens 1991). There are no large ceremonial centres, although the region may have been included in the hinterland to

Cahokia (with communication probably by river), and few status distinctions were apparent from burials here (Goldstein 1981).

Palaeodietary reconstruction with carbon isotopes

Plants have evolved more than one photosynthetic pathway, resulting in differences in their stable isotope ratios when analysed ($^{13}C/^{12}C$ reported as $\delta^{13}C$ values). Most of the flora of temperate North America utilise a C_3 photosynthetic pathway (Calvin and Benson 1948), including almost all of the seed plants of the indigenous Eastern Agricultural Complex. Consumption of these native plants – and of animals which have fed on them – will result in $\delta^{13}C$ values of around -20-22‰ in human bone collagen. Maize however, originally domesticated in Mexico, is a tropical plant with a C_4 photosynthetic pathway (Hatch and Slack 1966, 1967), which results in more enriched consumer bone collagen $\delta^{13}C$ values of -16‰ to -9‰, with an average -12‰.

This difference in $\delta^{13}C$ values means bone can be used to identify dietary change, and the presence of increasing amounts of C_4 maize in the diet (van der Merwe and Vogel 1978). $\delta^{13}C$ ratios may also increase if native amaranth is eaten (Schwarcz, et al. 1985), however archaeobotanical study reveals no evidence for amaranth consumption in this region (Asch and Asch 1985). Marine seafoods in the diet can also affect the carbon isotope signature (Sealy and van der Merwe 1988), but Illinois is a very long way from the sea. Bone collagen is rich in protein, and may underestimate maize consumption as the maize plant has a low protein content. Bone carbonate (apatite) better represents different dietary components (Krueger & Sullivan 1984), however it is prone to diagenesis, and there is no agreed way to identify and rectify such diagenetic contamination (Koch et al. 1997). In this study a diet of entirely C_3 resources is expected to produce an isotopic value of around -21‰, whilst carbon isotope values of -18.5‰ or more enriched are interpreted as evidence of maize being included as a notable component of diet (van der Merwe and Vogel 1978).

Skeletal samples and archaeological sites

The Joe Gay, Lawrence Gay, and Yokem sites are located in the Sny Bottom region of the Central Mississippi Valley in west-central Illinois, with Knight further to the south. Schild is in the neighbouring Lower Illinois Valley (Figure 1). Tables 1-4 provide assigned sex, age and $\delta^{13}C$ value for each individual, organised by period and site. Age ranges: Y = 15-35, M = 35-50, O = 50+.

The Lawrence Gay and Joe Gay burials probably originate from the same community (Della Cook, pers. comm.). The Lawrence Gay mound group (Cook 1976; Cook and Farnsworth 1981; Perino 1986) is located in the river bottoms. The burials recovered are believed to represent a sub-section of the population, individuals of high status: only the central log tombs were excavated, and, contrary to the Joe Gay site, females and juveniles were differentially distributed across the mounds, and denied access to some aspects of the mortuary program (Cook 1976; Cook and Farnsworth 1981). Interment in floodplain sites may generally have been more restricted than was burial in bluff-top sites (Buikstra 1976; Bullington 1988). The Middle Woodland components of these two sites therefore present a highly unusual opportunity to compare the diets of high and low status members of probably the same, contemporaneous group.

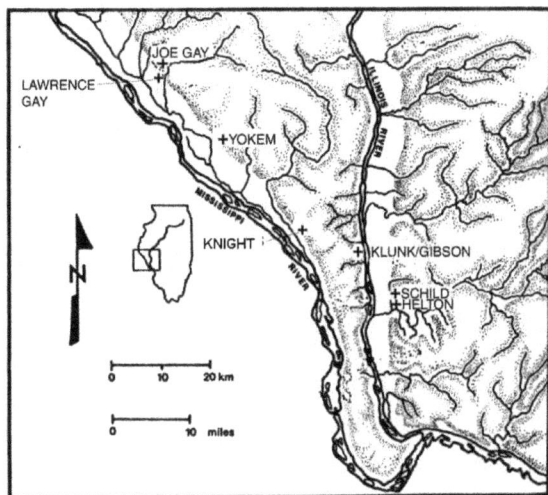

Figure 1. Map of west-central Illinois showing location of sites in this study. The state of Illinois is inset. In the CMVI Ledders is immediately south of Knight; in the LIV Koster Mounds are located between Schild and Helton. The American Bottom region (not shown here) is roughly 50km downriver south of the point where the rivers meet

The Joe Gay mounds on the adjacent blufftop had Middle and Late Woodland components (Cook 1976; Perino 2006c; Perino 2006b). Radiocarbon dating has been attempted on Late Woodland skeletal material, but unfortunately was problematic, probably due to radon contamination (Tainter 1977). Based on the presence of Whitehall pottery and corner- and side-notched points, the Late Woodland burials date from AD 600-800 (Conner 1984; Cook 1976; Perino 2006c). The Late Woodland burials were subdivided into early and late Late Woodland phases, identified by soil formation at intermediate zones in the mound fill of mounds 3 and 4. Selection of burials was carefully limited to only include individuals that the excavators felt could safely be assigned to the early or late Late Woodland (although this unfortunately reduced the number of burials available for analysis). Preservation was not particularly good at either site, and many bones had previously been stabilised with polyvinyl acetate (PVA).

The bluff top Knight mounds have Middle and early Late Woodland components, with an unexcavated associated village; human remains were dated from grave goods to AD 150-400 (Asch 1976; Griffin, et al. 1970; Wilkinson

1971). Few of the burials were curated, and those remaining date mostly from the early Late Woodland. Middle Woodland deposits were identified from grave goods. On the basis of ceramic continuity from the preceding Middle Woodland deposits, the excavators considered the Late Woodland material to date from very early in the Late Woodland, possibly as early as AD 400 (Griffin, et al. 1970). Bone preservation was good.

The Yokem mounds are located on a bluff on the eastern side of the Mississippi River. The site included 'middle Late Woodland' material (named by the excavator the "Yokem" phase) radiocarbon dated to AD 700-800, a late Late Woodland phase probably from the 11th-century AD, and Mississippian material radiocarbon dated to around AD 1200-1300; there was also a Late Woodland village which was not excavated. A Late Woodland village area was noted at Yokem, but was not excavated (Perino 1971b; Perino 2006a; Connor 1984). The "Yokem" phase skeletal material was poorly preserved and not suitable for analysis, but preservation of the more recent burials was good to excellent.

Schild in the LIV also had late Late Woodland and Mississippian components, the Mississippian burials being subdivided into the earlier Knoll A, and subsequent Knoll B (Perino 1971a; Perino 1973). Radiocarbon dates suggest that both temporal phases at Schild are a little earlier in date than the equivalent remains at Yokem, with the late Late Woodland remains dating from around AD 875 (Connor 1990; Droessler 1981), and Mississippian era material from around AD 1065 (Perino 1971a). A Late Woodland village was partially excavated at Schild, but there is no known Mississippian habitation site that could have supported the large burial population recovered from that period. Compared with the stark evidence for status distinctions known from Mississippian Cahokia and Dickson Mounds, there was relatively little differentiation at Schild (Goldstein 1980, 1981; Perino 1971a). Studies of both metric and non-metric cranial traits indicate probable genetic continuity from the Late Woodland (Droessler 1981). Bone preservation was excellent.

Materials and methods

Approximately 2g of cortical bone was removed from a rib from each burial, which was then cleaned ultrasonically. Some bones had previously been stabilised with PVA, and this was removed by pre-treatment with two 24-hour cycles in acetone (Moore, et al. 1989). Extraction of collagen from the bone was by slow demineralization in dilute HCl in sealed tubes over 14 days, except for some especially friable bone from which collagen was extracted with EDTA (Tuross, et al. 1988). Collagen extraction was followed by 24 hours in NaOH to remove humic components (Schoeninger, et al. 1989; Schwarcz and Schoeninger 1991), and the resulting fraction dried in an oven.

Collagen samples were converted to CO_2 and nitrogen gases by combustion in a Carlo-Erba elemental analyzer, followed immediately by mass spectroscopy with a Finnegan Delta Plus mass spectrometer, coupled to the elemental analyzer through a continuous flow interface. Pee Dee Belemnite limestone (PDB) served as the standard for comparison. Most biological material is less enriched in ^{13}C than PDB, and so carbon isotope values tend to be negative. The margin of error is approx 0.2‰; each run of samples was interspersed with the standard, sulfanilamide, for comparison.

Preservation was assessed by calculating % yield of collagen from each bone; a yield below 3% was deemed unreliable (Ambrose 1990; van Klinken 1999). The C:N ratio was also calculated during sample combustion, with a range of 2.9-3.6 being acceptable (Ambrose 1990; DeNiro 1985).

Results and Discussion

Middle Woodland

From the Middle Woodland sites the mean δ^{13}C ratio at Lawrence Gay was -20.8‰, at Joe Gay -20.6‰, and at Knight -21.0‰, with little variation at any site (Table 1). This indicates that diet was based overwhelmingly on indigenous C_3 plant species of the Eastern Agricultural Complex. These ratios are similar to those previously reported for sites in the neighbouring LIV, where Gibson produced an average of -21.0‰ (Bender, et al. 1981; van der Merwe and Vogel 1978), and Pete Klunk -20.6‰ (Schober 1998), and to results from elsewhere in the Eastern Woodlands (Buikstra 1992). Although maize is known in the archaeobotanical record from the Middle Woodland (Conard, et al. 1984), only traces have been recorded, so the lack of evidence for maize consumption at Middle Woodland Joe Gay, Lawrence Gay, and Knight is not unexpected.

There is little variation between Joe Gay and high status Lawrence Gay burials in δ^{13}C ratios (two-tailed Student's *t*-test *p*=0.1), so no evidence that high status individuals had greater access to maize or protein. It should be remembered, however, that protein-rich bone collagen probably under-represents maize consumption (since maize is low in protein): if maize consumption was low level and/or occasional (e.g. for rituals), differences might be so small as to fall within the range of instrument precision, and so not be apparent or statistically significant.

Late Woodland

It is in the Late Woodland that we would expect to see the first evidence for maize consumption, with carbon isotope ratios becoming more enriched. Most previous studies have put the date at AD 800 and after, in the late Late Woodland (e.g. Buikstra 1992). It is therefore surprising that early Late Woodland Knight should produce a mean carbon isotope ratio of -19.7‰, over 1‰ more enriched than the average for the Middle Woodland burials (one-tailed Student's *t*-test *p*=0.09), while at Joe Gay the mean -19.5‰ is nearly 1‰ more enriched than

previously (*p*=0.06) (Table 2). Of particular note, however, is the observation that at both sites this average value disguises several much more enriched δ^{13}C ratios, indicating substantial maize consumption in those individuals. At Joe Gay, two burials produced ratios of -16.6‰ and -15.9‰, while at Knight three burials produced ratios of -16.4‰, -14.4‰, -13.4‰. Similar isotopic evidence for maize consumption at Koster .

Table 1. Middle Woodland

Burial	Sex	Age	δ^{13}C ‰	C:N ratio	Extraction yield %
Lawrence	**Gay**	**N= 11**			
LG 1-4	M	Y	-20.2	3.55	1.2*
LG 2-4	M	M	-21.0	3.26	7.0*
LG 2-7	F	M	-21.5	3.52	5.4
LG 4-1	F	Y	-20.9	3.29	4.5
LG 4-2	M	O	-21.0	3.20	11.4
LG 4-3A	F	M	-20.2	3.26	3.2
LG 4-3B	M	O	-21.1	3.23	5.6
LG 5-1	M	O	-20.7	3.15	18.8
LG 5-3	M	M	-21.0	3.21	19.9
LG 7-3	M	M	-20.2	3.20	20.9
LG 10-1	M	M	-20.9	3.27	2.0*
Mean			-20.8	-	-
±1 SD			± 0.4	-	-
Joe Gay	**N**	**=13**			
JG 3-7F	M	M	-20.3	3.27	12.3
JG3-8B	F	M	-20.4	3.45	4.8*
JG 3-17A	M	Y	-21.1	3.31	10.8
JG 3-25	M	M	-20.9	3.21	20.2
JG 4-13	F	M	-20.5	3.27	10.5*
JG 4-21	M	M	-20.2	3.61	2.9
JG 4-32	F	M	-20.3	3.24	8.3*
JG 4-40	F	Y	-20.6	3.21	7.3
JG 4-54A	M	O	-20.4	3.45	2.7*
JG 4-54B	M	M	-20.4	3.43	2.7*
JG 4-61	F	Y	-20.8	3.16	3.4
JG 6-24B	F	Y	-20.7	3.17	4.4*
JG 6-24C	M	Y	-20.6	3.42	1.9*
Mean			-20.6	-	-
±1 SD			± 0.3	-	-
Knight	**N**	**=7**			
Kn 2-10	M		-20.6	3.15	-
Kn 2-74	M		-20.6	3.26	-
Kn 2-77	M		-20.5	3.22	-
Kn 3-17	M		-23.1	4.03	-
Kn 4-3	M		-20.2	3.10	-
Kn 4-5	M		-21.4	3.27	-
Kn 16-8	F		-20.6	3.29	-
Mean			-21.0	-	-
±1 SD			± 1.0	-	-

* = Mechanical loss during extraction.

Table 2. Early Late Woodland

Burial	Sex	Age	δ¹³C ‰	C:N	Extraction
Joe Gay	**N**	**=9**			
JG 3-10	F	Y	-16.6	3.24	15.3
JG 3-23	F	Y	-20.6	3.42	4.9
JG 3-34	M	Y	-20.6	3.31	14.3
JG 4-7	M	Y	-20.2	3.37	10.2
JG 4-22	F	O	-20.1	3.38	3.3
JG 4-33	M	Y	-15.9	3.24	10.6
JG 4-44	M	M	-20.5	3.51	0.8*
JG 4-48	F	M	-20.2	3.34	7.2
JG 4-71	F	Y	-20.7	3.23	12.9
Mean			-19.5	-	-
±1 SD			± 1.9	-	-
Knight	**N**	**=17**			
Kn 1-1	M		-21.6	3.28	-
Kn 1-8	M		-20.4	3.42	-
Kn 1-10	F		-21.5	4.36	-
Kn 1-14	F		-20.7	3.20	-
Kn 1-15	M		-20.5	3.24	-
Kn 1-17	M		-20.5	3.21	-
Kn 1-18	M		-20.7	3.54	-
Kn 1-20	F		-20.8	3.37	-
Kn 2-7	M		-20.7	3.08	-
Kn 2-9	F		-20.5	3.16	-
Kn 2-69	M		-21.3	4.14	-
Kn 2-71	M		-13.4	3.29	-
Kn 2-72	M		-14.4	3.23	-
Kn 3-10	F		-21.1	3.28	-
Kn 3-14	F		-20.9	3.38	-
Kn 17-1	F		-16.4	3.34	-
Kn 21-1	F		-20.5	3.17	-
Mean			-19.7	-	-
±1 SD			± 2.5	-	-

* = Mechanical loss during extraction.

Table 3. Late Late Woodland

Burial	Sex	Age	δ¹³C ‰	C:N ratio	Extraction
Joe Gay	**N**	**=15**			
JG 3-1	F	Y	-15.8	3.35	11.8
JG 3-12	F	Y	-20.3	3.34	17.1
JG 3-14	M	M	-20.3	3.30	17.4
JG 3-22	M	O	-21.0	3.23	19.2
JG 3-27A	M	O	-20.0	3.24	11.5
JG 3-30	F	Y	-20.5	3.28	7.2
JG 3-31	M	Y	-18.2	3.45	4.7
JG 4-3	M	Y	-15.4	3.26	14.3
JG 4-5	M	Y	-16.1	3.30	11.6
JG 4-12	F	Y	-20.5	3.23	16.3
JG 4-20	M	M	-16.2	3.28	16.3
JG 4-29	F	M	-20.3	3.24	4.5
JG 4-31	M	Y	-20.1	3.20	9.5
JG 4-34	F	O	-20.5	3.21	19.8
JG 4-59	F	O	-21.1	3.52	5.3
Mean			-19.1		
±1 SD			± 4.5		

Burial	Sex	Age	δ¹³C ‰	C:N	Extraction
Yokem	**N**	**=18**			
YO 3-3	F	M	-19.9	3.42	8.2
YO 3-7	F	O	-19.4	3.29	17.3
YO 3-11	F	M	-20.9	3.33	18.9
YO 3-12A	F	Y	-11.5	3.29	20.5
YO 3-16	M	M	-20.2	3.37	10.4
YO 3-49	M	Y	-12.6	3.28	20.3
YO 3-85	M	O	-19.8	3.35	11.0
YO 4-1	F	M	-13.2	3.34	17.7
YO 4-6	M	M	-18.0	3.96	22.0
YO 4-7	F	Y	-11.8	3.30	14.3
YO 4-9	M	O	-14.6	3.29	16.0
YO 4-16	F	O	-14.9	3.31	22.5
YO 4-32	M	Y	-13.8	3.28	18.8
YO 5-7	F	O	-14.4	3.33	15.5
YO 5-11	F	M	-13.8	3.30	11.6
YO 5-14	F	Y	-13.3	3.28	22.9
YO 5-20	M	O	-15.3	3.54	4.9
YO 5-39	M	M	-13.8	3.24	14.2
Mean			-15.6		
±1 SD			± 3.2		
Schild	**N**	**=39**			
S 1-7†	M	M	-19.9	-	-
S 1-11	F	O	-18.5	3.30	21.8
S 1-13	F	O	-20.1	3.40	17.7
S 1-16†	F	Y	-19.9	-	
S 1-17†	M	M	-18.2	-	
S 1-23	M	O	-20.0	3.36	13.9
S 1-24A	F	O	-18.9	3.57	10.1
S 1-28†	F	Y	-19.1	-	
S 1-34	M	O	-19.8	3.37	21.0
S 1-40	F	M	-20.2	3.27	22.9
S 1-62	M	O	-19.6	3.36	20.9
S 1-71	F	M	-19.4	3.31	18.3
S 1-73	M	M	-18.7	3.28	23.6
S 1-74	F	Y	-18.2	3.27	19.4
S 2-11A†	M	Y	-18.3	-	
S 2-15†	M	Y	-19.1	-	
S 2-31	M?	Y?	-19.2	3.34	19.8
S 3-2C	M	Y?	-13.9	3.31	15.0
S 3-6†	M	Y	-19.2	-	-
S 3-16†	M	M	-19.1	-	
S 3-17A†	F	M	-19.4	-	-
S 3-20A†	M	M	-20.0	-	
S 3-21	F	Y	-20.6	3.35	10.3
S 3-24†	F	Y	-20.4	-	
S 3-25†	M	M	-20.4	-	
S 3-27†	M	Y	-20.6	-	
S 3-28†	F	Y	-20.5	-	
S 3-32A†	F	Y	-20.2	-	
S 9-7†	F	Y	-20.4	-	-
S 9-13	F	O	-21.1	3.30	20.5
S 9-17A†	F	Y	-20.3	-	
S 9-24	M	O	-20.9	3.27	15.9
S 9-28	M	M	-20.4	3.27	13.5
S 9-28A	F	M	-20.6	3.21	21.1
S 9-32	F	M	-20.5	3.28	17.1
S 9-34	M	M	-20.1	3.33	18.6

87

Table 3: contd.

S 9-37A†	F	Y	-20.3		-
S 9-38	M	Y	-20.4	3.29	6.8
S 9-40†	F	Y	-20.6		-
Mean			-19.7		
±1 SD			± 1.2		

† Data from Schober 1998. Extraction yields and C:N ratios were not published, but were described as within the acceptable range defined in the Results section here.

(a) Knight

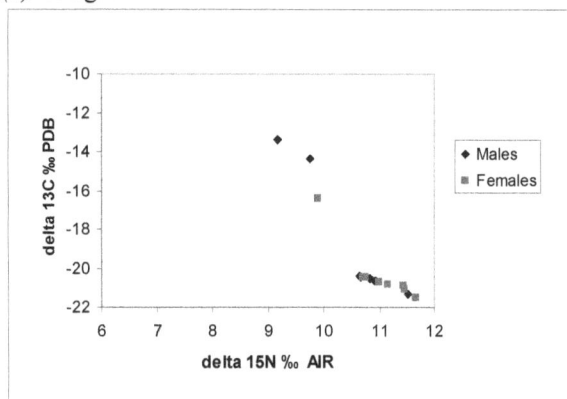

(b) Joe Gay

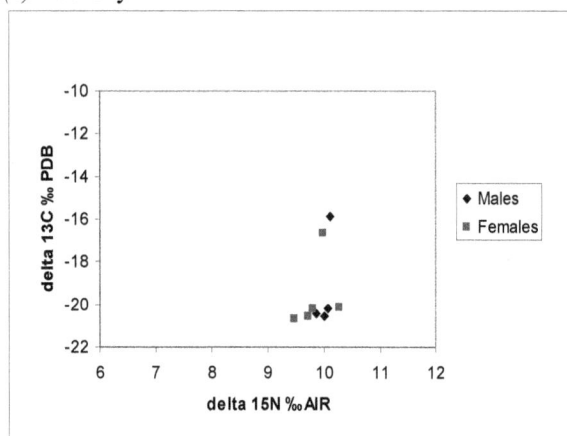

(c) Koster Mounds*

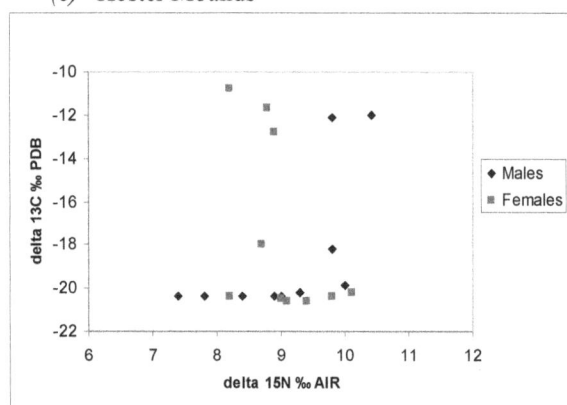

*Data from Schober 1998. *n*=20 village does include late Late Woodland and Mississippian material) (Griffin, et al. 1970).

Figure 2. Scatterplots showing bi-modal distribution of carbon isotope ratios in the early Late Woodland.

Mounds was reported in the neighbouring LIV, where the mean -18.0‰ included a cluster of values around -12‰ (Schober 1998). When the data is plotted out, it is also noticeable that the distribution of $\delta^{13}C$ ratios is bi-modal at all three sites (Figure 2).

The excavators of Knight believed the early Late Woodland remains to date from as early as AD 400 (Griffin, et al. 1970). The maize consumers could be intrusive later burials, but there was no late Late Woodland mortuary component recognized at Knight mounds (although the nearby village does include late Late Woodland and Mississippian material) (Griffin, et al. 1970). Joe Gay had both early and late Late Woodland interments, and it is possible that these maize consumers have been wrongly labelled as 'early' Late Woodland, however burial selection was carefully limited to only include individuals that the excavators felt could safely be assigned to the early Late Woodland. Radiocarbon dates at LIV Koster Mounds showed no relationship with bone collagen $\delta^{13}C$ values (Schober 1998), undermining suggestions that the more enriched $\delta^{13}C$ ratios came from the end of the early Late Woodland. The recording in this area of maize consumption at such an early date is not entirely surprising, however, as there is a concentration of archaeobotanical finds of maize in the early Late Woodland from around the CMVI (Simon 2000). We may be seeing an emerging trend in this region towards intensification in maize use earlier than in other parts of the Eastern Woodlands.

The trend for increased maize consumption continues in the late Late Woodland (Table 3). At Joe Gay the average -19.1‰ now disguises an even larger number of more enriched carbon isotope ratios, whilst at Yokem the majority of individuals were eating maize, producing a mean -15.6‰. Previously reported are averages from Ledders -17.4‰ and Helton -17.5‰ (Buikstra, et al. 1987, Buikstra personal communication). Schild in the LIV stands out with little variation around a mean of 19.6‰, showing scant evidence for maize consumption compared with the other late Late Woodland sites. This may be because Schild is a little earlier in date (Connor 1990; Droessler 1981), but there was clear evidence for maize consumption previously in the LIV, at early Late Woodland Koster Mounds which are not far away, so the evidence for only minor maize consumption at Schild is enigmatic, and again points to variability in uptake of the new crop.

The late Late Woodland results are again notable for so much intra-community variability in carbon isotope ratios that the data is bi-modal when plotted out, except at Ledders where the distribution was continuous (Figure 3). The carbon isotope ratios encompass a dietary maize component varying from near zero, to a very high percentage, apparently within the same population.

(a) Joe Gay

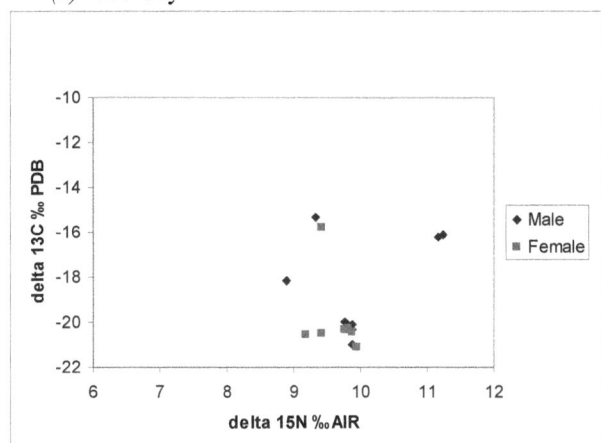

Table 4.

(b) Yokem

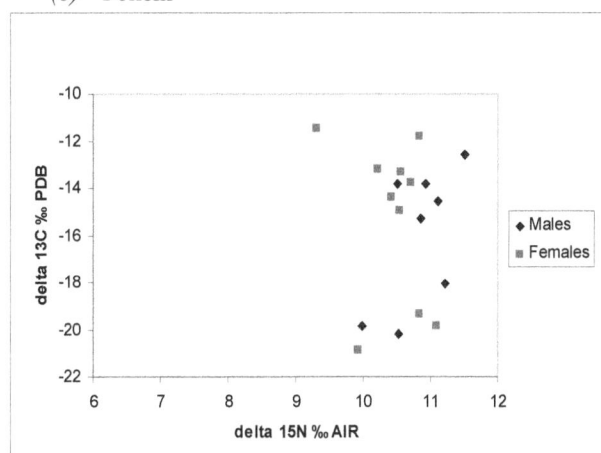

(c) Helton Mound 47*

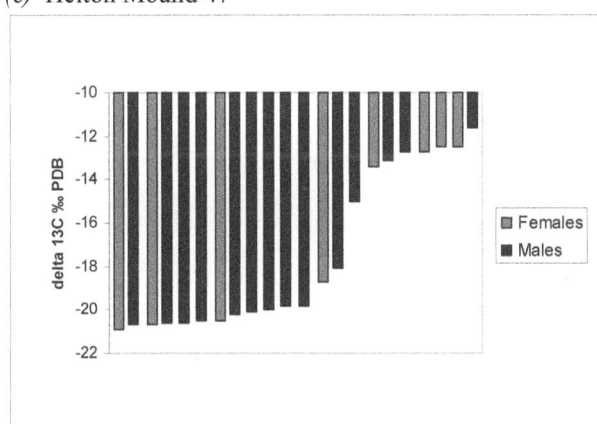

* =late Late Woodland/"emergent Mississippian" Data from Buikstra personal communication 2002; only carbon isotope ratios available.

Figure 3. Scatterplots showing bi-modal distribution of carbon isotope ratios in the late Late Woodland period.

Burial	Sex	Age	$\delta^{13}C$ ‰	C:N ratio	Extraction yield %
Yokem N=18					
YO 1-3	F	O	-20.6	3.56	7.5
YO 1-7	F	O	-13.8	3.34	11.7
YO 1-30	M	Y	-13.8	3.30	12.8
YO 1-34	M	Y	-15.2	3.85	6.3
YO 1-36	M	O	-15.8	3.48	15.2
YO 1-43	M	M	-15.5	3.56	9.5
YO 1-44	F	M	-14.5	3.29	15.2
YO 1-52	M	O	-14.3	3.24	16.8
YO 1-55	F	O	-12.5	3.20	18.5
YO 1-57	F	Y	-12.3	3.20	14.2
YO 1-64	M	O	-12.8	3.27	11.6
YO 2-2	F	Y	-14.0	3.36	11.2
YO 2-6	M	M	-14.3	3.48	7.6
YO 2-7	F	O	-13.8	3.26	19.4
YO 2-13	M	Y	-15.1	3.23	10.9
YO 3-97	M	Y	-13.3	3.42	7.2
YO 3-104	F	M	-12.8	3.17	17.9
YO 3-115	F	M	-18.3	3.19	10.4
Mean			-14.6		
±1 SD			± 2.1		
Schild Knoll A N=24					
S A-42	M	M	-21.2	3.35	11.9
S A-45	M	O	-20.8	3.54	4.7
S A-48	M	M	-21.5	3.28	16.1
S A-49	F	M	-13.6	3.26	21.5
S A-56	F	O	-12.0	3.29	15.9
S A-62	F	Y	-13.7	3.20	17.1
S A-65†	F	Y	-13.2	-	
S A-66A†	F	Y	-14.8	-	
S A-69A†	F	Y	-12.3	-	
S A-74	F	M	-12.6	3.24	17.9
S A-76	F	M	-16.8	3.24	12.1
S A-82	F	M	-15.5	3.37	9.7
S A-83A	M	M	-13.7	3.27	15.4
S A-85†	F	Y	-10.4	-	
S A-101A†	M	Y	-12.8	-	
S A-111	F	Y	-14.4	3.24	17.8
S A-136	M	O	-14.6	3.52	8.3
S A-137	M	M	-16.1	3.21	16.8
S A-149†	M	Y	-15.3	-	
S A-153†	M	Y	-20.4	-	
S A-154	M	Y	-14.8	3.59	5.7
S A-156	M	Y	-13.8	3.26	18.3
S A-157†	F	Y	-15.3	-	
S A-158	F	O	-13.0	3.22	20.8
Mean			-15.1		
±1 SD			± 3.0		
Schild Knoll B N=14					
S B-165	M	M	-11.3	3.52	9.7
S B-166†	F	Y	-12.3	-	
S B-175A	F	M	-11.6	3.28	16.4
S B-194†	M	Y	-13.6	-	
S B-201	F	Y	-13.7	3.17	20.9
S B-207	M	O	-12.9	3.24	21.7
S B-217	F	Y	-13.4	3.26	7.1
S B-219	F	M	-14.6	3.19	20.9
S B-221A†	M	Y	-13.1	-	
S B-235	M	Y	-13.6	3.24	13.1

Table 4 contd.

S B-250	M	M	-12.8	3.35	10.1
S B-263†	F	M	-12.1		-
S B-285	F	O	-15.3	3.40	8.9
S B-296†	M	Y	-11.4		-
Mean			-13.0		
±1 SD			± 1.2		

† Data from Schober 1998. Extraction yields and C:N ratios were not published, but were described as within the acceptable range defined in the Results section here.

Mississippian

By Mississippian times the average carbon ratios are yet more enriched, -14.6‰ at Yokem and at Schild -15.1‰ for earlier Knoll A, and -13.0‰ from Knoll B (Table 4). These values reflect the substantial contribution of maize to the diet that would be expected for Mississippian people (although these values are not as enriched as some of the ratios reported for Fort Ancient populations to the east in Ohio). Even now, however, there were a few individuals who were still eating predominately C_3 plants (although with a small maize contribution) leading to a bi-modal distribution at Yokem and Schild Knoll A, but now the balance has changed with most community members eating a substantial quantity maize (Figure 4). The sample from Schild Knoll B is smaller, but the mean -13.0‰ indicates that the entire community was now consuming a considerable amount of maize.

Intra-community variation

Age does not correlate with diet; burials were divided into rough age groups (young, middle-aged, and old) but only minor age differences were found, within the margin of instrument error, and are not considered noteworthy. Sex differences in carbon isotope ratios were recorded once only, at early Mississippian Schild (Knoll A), where females had eaten much more maize than males (two-tailed Student's *t*-test *p*= 0.015) (Table 4). Biomechanical evidence from the LIV at this time suggests women apparently had a greater role in plant food production (Bridges, Blitz, and Solano 2000). If they were 'early adopters' of the new crop this is a possible explanation for their higher consumption, although one would presume that food would not vary much throughout the household, a unit that would usually include male members. Alternatively, it could reflect some kind of dietary discrimination against females, although we do not know how maize was perceived by Mississippian groups. At Cahokia Mound 72, sacrificed female slaves had consumed markedly more maize than both the male slaves buried alongside them, and than the other social groups interred in the mound, and had suffered poorer health generally (Buikstra, Rose, and Milner 1994), but slaves in a great ceremonial centre are not perhaps the best analogue for Schild. The finding is enigmatic, as sex differences are not observed at other sites.

The one significant and consistent difference was the bi-modal pattern in crop consumption as maize is adopted

(Figs. 3, 4, & 5). As mentioned, neither sex nor age differences correlate significantly with variation in maize consumption, nor is there much archaeological evidence for status differences at these sites. The maize eaters are unlikely to be immigrants (Schober 1998), since, particularly at the earlier sites, the nearest groups consuming much maize were hundreds of miles away, and even in the late Late Woodland there is no archaeological or biodistance evidence for immigration at this time (Conner 1990; Konigsberg 1990).

It is possible that the evidence is not pointing to true intra-community variation in diet, but that the two groups are in fact separated by a small degree of temporal variation between the burials not picked up by the excavators. Fluoride dating was employed to test this hypothesis, a relative dating method which has successfully been used to separate late prehistoric burials from closely temporally-related contexts (Schurr 1989; Ezzo 1992). The longer a bone is buried, the more fluoride it will absorb from groundwater. Because the material dated (bones) must come from the same burial environment, such dating was only possible within early and late Late Woodland components at Joe Gay and late Late Woodland at Yokem. Fluoride dating showed no correlation between relative age and maize consumption in any temporal component at either site. This indicates that the bi-modal pattern of crop consumption is real, and that the two groups are indeed contemporary. This finding is supported by the persistence of the bi-modal pattern of crop consumption at several sites, over several hundred years from early Late Woodland to early Mississippian.

The ratio of maize consumers to non-maize consumers changes over time, but the presence of two discrete consumption patterns does not. Why might this be? Apparently indigenous crops continued to suffice for many households, possibly a consequence of the lower investment of time and energy required to produce them. If, however, maize was better suited to intensification than local crops, perhaps some households had a greater need to exploit this? Bigger families, for example, had more mouths to feed, but they also had the additional labour needed to grow maize. Some families were perhaps more risk averse, and maize-growing may have been a more reliable option. There is evidence of greater population density through the Late Woodland: as communities grew, some gardens or field plots must have been smaller, or on poorer soil than others, perhaps making more intensive maize growing a better strategy for those households. If surpluses were becoming important at this time, perhaps for feasting or communal storage against shortage, it may have been easier to satisfy such requirements with maize as it is easier to store. Some households may have had a bigger 'bill' to settle, or were less able to satisfy such demands from their existing production. Finally, it should not be forgotten that maize is tastier and more versatile than bland native crops, a consideration which may have played a role in deciding between the two crops.

Conclusions

This study has confirmed that maize was not eaten in any quantity in the Middle Woodland period. There are obvious status differences between burials at that time, and early maize use has been linked to ritual or status-related activity, but there were no differences m between high and low status burials.

Maize consumption appears to become popular earlier in this region than elsewhere in the Eastern Woodlands. Maize was not believed to become common until after AD 800, however more enriched carbon isotope ratios are repeatedly evident in early Late Woodland burials in this area, at the CMVI Knight site possibly as early as AD 400. If correct, this would make Knight one of the very earliest dates for substantial maize consumption in the Eastern Woodlands.

The adoption of maize as a staple crop is not quick – as has been speculated – but occurs gradually over several hundred years. What was particularly noticeable, and unexpected, was the variability in uptake within the community, a characteristic which is masked by the average isotopic values. The distribution of carbon isotope ratios is bi-modal when plotted out, indicating that some sections of the community continued to eat a diet based entirely on indigenous C_3 plants long after others had incorporated substantial quantities of maize into their diets. The balance between the two groups gradually changes over time, with the maize-eaters eventually predominating, but the bi-modal pattern is persistent. There is also variation in maize uptake between sites of similar time period, Schild in this study being notable for the late rise in maize consumption in comparison with neighbouring sites.

Acknowledgments

This work was part of a PhD dissertation at Rutgers, the State University of New Jersey (USA). Thanks are expressed to; Mark R. Schurr (University of Notre Dame) for supervision of laboratory work, and access to ND's Reynier Laboratory for Biocultural Studies and Center for Environmental Science & Technology, Della Cook at Indiana University-Bloomington for access to the Gay mounds, Yokem, and Schild skeletal material, Smithsonian Institution, Washington D.C. for access to the Knight material, and Jane Buikstra for sharing Helton & Ledders isotope data. Grant sponsorship: National Science Foundation (USA), Rutgers University Graduate School Excellence Fellowships, Rutgers University Anthropology Department Antoinette Bigel Fund.

Literature Cited

Ambrose SH (1990) Preparation and characterization of bone and tooth collagen for isotopic analysis. Journal of Archaeological Science 17: 431-51.

Asch DL (1976) The Middle Woodland Population of the Lower Illinois Valley: a Study in Paleodemographic Methods. Evanston, IL, Northwestern University Archeological Program. Scientific Papers no. 1.

Asch DL and Asch NB (1985) Prehistoric plant cultivation in West-Central Illinois. In RI Ford (ed.): Prehistoric Food Production in North America. Ann Arbor, MI: Museum of Anthropology, University of Michigan. Anthropological Papers 75: 149-204.

Bender MM, Baerreis DA and RL Steventon (1981) Further light on carbon isotopes and Hopewell agriculture. American Antiquity 46: 346-53.

Braun DP (1979) Illinois Hopewell burial practices and social organization: a reexamination of the Klunk-Gibson mound group. In DS Brose and N Greber (eds.): Hopewell Archaeology: the Chillicothe Conference. Kent, OH: Kent State University Press; 66-79

Bridges PS, Blitz JH and MC Solano (2000) Changes in long bone diaphyseal strength with horticultural intensification in west-central Illinois. American Journal of Physical Anthropology 112: 217-38.

Brown JA (1979) Charnel houses and mortuary crypts: disposal of the dead in the Middle Woodland period. In DS Brose and N Greber (eds.): Hopewell Archaeology: the Chillicothe Conference Kent, OH: Kent State University Press; 211-19.

Buikstra JE (1976) Hopewell in the Lower Illinois Valley: a Regional Approach to the Study of Human Biological Variability and Prehistoric Behavior. Chicago: Northwestern University Archaeological Program, Scientific Papers 2.

Buikstra JE (1984) The Lower Illinois River region: a prehistoric context for the study of ancient diet and health. In M Cohen and G Armelagos (eds.): Paleopathology at the Origins of Agriculture. Orlando, FL: Academic Press; 217-35.

Buikstra JE (1992) Diet and disease in late prehistory. In JW Verano and DH Ubelaker (eds.): Disease and Demography in the Americas. Washington, DC: Smithsonian Institution Press; 87-101.

Buikstra JE, Bullington J, Charles DK, Cook DC, Frankenberg SR, Konigsberg L, Lambert JB and L Xue (1987) Diet, demography, and the development of horticulture. In W Keegan (ed.): Emergent Horticultural Economies of the Eastern Woodlands. Carbondale, IL: Center for Archaeological Investigations, Southern Illinois University at Carbondale, Occasional Paper no. 7; 67-85.

Buikstra JE, Rose JC and GR Milner (1994) A carbon isotopic perspective on dietary variation in late prehistoric western Illinois. In W. Green (ed.): Agricultural Origins and Development in the Midcontinent Iowa City, IA: University of Iowa; 155-70.

Bullington J (1988) Middle Woodland mound structure: social implications and regional context. In DK Charles, SR Leigh and JE Buikstra (eds.): The Archaic and Woodland Cemetaries in the Lower Illinois Valley Kampsville, IL: Center for American Archeology; 218-41.

Calvin M and Benson AA (1948) The path of carbon in photosynthesis. Science 107: 476-80.

Chapman, J and Crites, GD (1987) Evidence for early maize (*Zea mays*) from the Icehouse Bottom site, Tennessee. American Antiquity 52:362-4.

Charles DK (1995) Diachronic regional social dynamics: mortuary sites in the Illinois Valley/American Bottom region. In LA Beck (ed.): Regional Approaches to Mortuary Analysis. New York: Plenum Press; 77-99.

Conard N, Asch DL, Asch NB, Elmore D, Gove H, Rubin M, Brown JA, Wiant MD, Farnsworth KB and TG Cook (1984) Accelerator radiocarbon dating of evidence for prehistoric horticulture in Illinois. Nature 308: 443-446.

Conner MD (1984) Population Structure and Biological Variation in the Late Woodland of Westcentral Illinois. Unpublished Ph.D. thesis, University of Chicago.

Connor MD (1990) Population structure and skeletal variation in the Late Woodland of West-Central Illinois. American Journal of Physical Anthropology 82: 31-43.

Connor MD (1991) Summary and conclusions. In KA Atwell and MD Connor (eds.): The Kuhlman Mound Group and Late Woodland Mortuary Behavior in the Mississippi River Valley of West-Central Illinois. Kampsville, IL: Center for American Archaeology; 238-45.

Cook DC (1976) Pathologic States and Disease Process in Illinois Woodland Populations: an Epidemiologic Approach. Unpublished PhD thesis, University of Chicago.

Cook DC (1984) Subsistence and health in the Lower Illinois Valley: osteological evidence. In MN Cohen and GJ Armelagos (eds.): Paleopathology at the Origins of Agriculture. Orlando, FL: Academic Press; 237-69.

Cook DC and Farnsworth KB (1981) Clay funerary masks in Illinois Hopewell. Mid-Continental Journal of Archaeology 6: 3-15.

DeNiro MJ (1985) Postmortem preservation and alteration of in vivo bone collagen isotope ratios in relation to palaeodietary reconstruction. Nature 317: 806-9.

Droessler J (1981) Craniometry and Biological Distance: Biocultural Continuity and Change at the Late Woodland

- Mississippian Interface. Evanston, IL: Center for American Archeology.

Ezzo JA (1992) Refinement of the adult burial chronology of Grasshopper Pueblo, Arizona. Journal of Archaeological Science 19: 445-57.

Farnsworth KB, Emerson TE and R Miller Glenn (1991) Patterns of Late Woodland/Mississippian interaction in the Lower Illinois Valley drainage. In TE Emerson and R. Lewis (eds.): Cahokia and the Hinterlands. Urbana, IL: University of Illinois; 83-117.

Fritz GJ (1990) Multiple pathways to farming in Pre-Contact Eastern North America. Journal of World Prehistory 4: 387-435.

Gallagher JP (1989) Agricultural intensification and ridged-field cultivation in the prehistoric upper Midwest of North America. In DR Harris and GC Hillman (eds.): Foraging and Farming. London: Unwin Hyman; 572-84.

Gallagher JP (1992) Prehistoric field systems in the Upper Midwest. In WI Woods (ed.): Late Prehistoric Agriculture: Observations from the Midwest Springfield, IL: Illinois Historic Preservation Agency; 95-135

Goldstein LG (1980) Mississippian Mortuary Practices. A Case Study of Two Cemeteries in the Lower Illinois Valley. Evanston, IL: Northwestern Archaeological Program Scientific Papers 4.

Goldstein LG (1981) One-dimensional archaeology and multi-dimensional people: spatial organisation and mortuary analysis. In R. Chapman, I. Kinnes, and K. Randsborg (eds.): The Archaeology of Death. Cambridge: Cambridge University Press; 53-69.

Gremillion KJ (1996) Diffusion and adoption of crops in evolutionary perspective. Journal of Anthropological Archaeology 15: 183-204.

Gremillion KJ (2004) Seed processing and the origins of food production in Eastern North America. American Antiquity 69: 215-233

Griffin JB, Flanders RL and PF Titterington (1970). The Burial Complexes of the Knight and Norton Mounds in Illinois and Michigan. Ann Arbor, MI, Museum of Anthropology, University of Michigan. Memoirs no. 2.

Hart J (1999) Maize agriculture evolution in the Eastern Woodlands of North America: a Darwinian perspective. Journal of Archaeological Method and Theory 6: 137-180

Hastorf CA (1998) The cultural life of early domestic plant use. Antiquity 72: 773-782.

Hastorf CA and Johannessen S (1994) Becoming corn-eaters in prehistoric America. In S Johannessen and CA

Hastorf (eds.): Corn and Culture in the Prehistoric New World. Boulder, CO: Westview Press: 427-43.

Hatch MD and Slack CR (1966) Photosynthesis by sugarcane leaves. A new carboxylation reaction and the pathway of sugar formations. Biochemical Journal 101: 103-11.

Hatch MD and Slack CR (1967) Further studies on a new pathway of photosynthetic carbon dioxide fixation in sugarcane, and its occurrence in other species. Biochemical Journal 102: 417-22.

Johannessen S (1993) Farmers of the Late Woodland. In CM Scarry (ed.): Foraging and Farming in the Eastern Woodlands. Gainesville: University Press of Florida: 57-77.

Koch PL, Tuross N and ML Fogel (1997) The effects of sample treatment and diagenesis on the isotopic integrity of carbonate in biogenic hydroxylapatite. Journal of Archaeological Science 24: 417-29.

Konigsberg LW (1990) Temporal aspects of biological distance: serial correlation and trend in a prehistoric skeletal lineage. American Journal of Physical Anthropology 82: 45-52.

Krueger HW and Sullivan CH (1984) Models for carbon isotope fractionation between diet and bone. In JE Turnland and PE Johnson (eds.): Stable Isotopes in Nutrition. American Chemical Society, Symposium Series no. 258: 205-22.

Lambert JB, Szpunar CB and JE Buikstra (1979) Chemical analysis of excavated human bone from Middle and Late Woodland sites. Archaeometry 21:115-29.

Moore KM, Murray ML and MJ Schoeninger (1989) Dietary reconstruction from bones treated with preservatives. Journal of Archaeological Science 16: 437-46.

Muller J and Stephens, JE (1991) Mississippian sociocultural adaptation. In TE Emerson and RB Lewis (eds.): Cahokia and the Hinterlands. Urbana, IL: University of Illinois; 297-310.

O'Brien, MJ (1986) Hopewell in the Lower Illinois River valley. Quarterly Review of Archaeology 7: 3-5.

O'Brien MJ (1987) Sedentism, population growth, and resource selection in the Woodland Midwest. Current Anthropology 28: 177-97.

O'Brien MJ and Pulliam CB (1996) Exploitation of floral resources. In MJ O'Brien (ed.): Middle and Late Woodland Subsistence and Ceramic Technology in the Central Mississippi River Valley. Springfield: Illinois State Museum, Reports of Investigations 52; 177-202.

Perino GR (1971a) The Mississippian component of the Schild site (no. 4), Greene County, Illinois. Illinois Archaeological Survey Bulletin 8: 1-148.

Perino GR (1971b) The Yokem site, Pike County, Illinois. Illinois Archaeological Survey Bulletin 8: 149-86.

Perino GR (1973) The Late Woodland component at the Schild Site, Greene County, Illinois. Illinois Archaeological Survey Bulletin 9.

Perino GR (1986) A Hopewell bone scepter. Central States Archaeological Journal (October): 357-60.

Perino, GR (2006a) Excavation of the Yokem site Late Woodland mounds, Pike County, Illinois. In KB Farnsworth and MD Wiant (eds.) Illinois Hopewell and Late Woodland Mounds: the Excavations of Gregory Perino 1950-1975. ITARP / University of Illinois Studies in Archaeology 4, Urbana, IL; 347-390.

Perino, GR (2006b) The 1970 Joe Gay mounds excavations, Pike County, Illinois. In KB Farnsworth and MD Wiant (eds.) Illinois Hopewell and Late Woodland Mounds: the Excavations of Gregory Perino 1950-1975. ITARP / University of Illinois Studies in Archaeology 4, Urbana, IL; 471-504.

Perino, GR (2006c) The 1970 Lawrence Gay mounds excavations, Pike County, Illinois. In KB Farnsworth and MD Wiant (eds.) Illinois Hopewell and Late Woodland Mounds: the Excavations of Gregory Perino 1950-1975. ITARP / University of Illinois Studies in Archaeology 4, Urbana, IL; 535-538.

Riley TJ, Walz GR, Bareis CJ, Fortier CJ and KE Parker (1994) Accelerator mass spectrometry (AMS) dates confirm early *Zea mays* in the Mississippi River valley. American Antiquity 59: 490-98.

Rose FR (in prep.) A fresh perspective: isotopic evidence for Illinois subsistence from the Mississippi River valley, and comparison with the Lower Illinois River valley. Midcontinental Journal of Archeology.

Schober T (1998) Reinvestigation of Maize Introduction in West-Central Illinois: a Stable Isotope Analysis of Bone Collagen and Apatite Carbonate from Late Archaic to Mississippian Times. M.A. thesis, Dept. of Anthropology, University of Illinois-Urbana Champaign.

Schoeninger MJ, Moore KM, Murray ML and JD Kingston (1989) Detection of bone preservation in archaeological and fossil samples. Applied Geochemistry 4: 281-92.

Schurr MR (1989) Fluoride dating of prehistoric bones by ion selective electrode. Journal of Archaeological Science 16: 265-70.

Schwarcz HP, Melbye J, Katzenberg MA, and M Knyf (1985) Stable isotopes in human skeletons of Southern Ontario: reconstructing palaeodiet. Journal of Archaeological Science 12: 187-206.

Schwarcz HP and Schoeninger MJ (1991) Stable isotope analysis in human nutritional ecology. Yearbook of Physical Anthropology 34: 283-321.

Sealy JC and van der Merwe NJ (1988) Social, spatial and chronological patterning in marine food use as determined by $\delta^{13}C$ measurements of Holocene human skeletons from the south-western Cape, South Africa. World Archaeology 20: 87-102.

Simms SR (1987) Behavioural Ecology and Hunter-Gatherer Foraging: an Example from the Great Basin. British Archaeological Reports, International Series 381.

Simon ML (2000) Regional variations in plant use strategies in the Midwest during the Late Woodland. In TE Emerson, DL McElrath and AC Fortier (eds.): Late Woodland Societies. Lincoln: University of Nebraska: 37-76.

Smith BD (1987) The independent domestication of indigenous seed-bearing plants in Eastern North America. In W Keegan (ed.): Emergent Horticultural Economies of the Eastern Woodlands. Carbondale: Center for Archaeological Investigations, Southern Illinois University Carbondale. Occasional Paper 7: 3-47.

Smith BD and Cowan CW (2003) Domesticated crop plants and the evolution of food production economies in Eastern North America. In PE Minnis (ed.): People and Plants in Eastern North America. Washington D.C.: Smithsonian; 105-125

Stothers DM and Bechtel SK (1987) Stable carbon isotope analysis: an inter-regional perspective. Archaeology of Eastern North America 15: 137-154.

Styles BW (1981) Faunal Exploitation and Resource Selection. Evanston: Northwestern University Archaeological Program. Scientific Papers no. 3.

Tainter JA (1977) Woodland social change in west-central Illinois. Mid-Continental Journal of Archaeology 2: 67-98.

Tuross N, Fogel ML and Hare PE (1988) Variability in the preservation of the isotopic composition of collagen from fossil bone. Geochimica et Cosmochimica Acta 52: 929-35.

van der Merwe NJ and Vogel JC (1978) $\delta^{13}C$ content of human collagen as a measure of prehistoric diet in woodland North American. Nature 276: 815-6.

van Klinken GJ (1999) Bone collagen quality indicators for palaeodietary and radiocarbon measurements. Journal of Archaeological Science 26: 687-95.

Wagner GE (1994) Corn in Eastern Woodlands late Prehistory. In S Johannessen and CA Hastorf (eds.): Corn and Culture in the Prehistoric New World. Boulder, CO: Westview Press: 335-346.

Watson PJ and Kennedy M (1991). The development of horticulture in the Eastern Woodlands of North America: women's role. In J Gero and M Conkey (eds.): Engendering Archaeology. Cambridge, MA: Blackwell.

Wilkinson RG (1971) Prehistoric Biological Relationships in the Great Lakes Region. Ann Arbor, MI, Museum of Anthropology, University of Michigan. Anthropological Papers no. 43.

Wymer DA (1993) Cultural change and subsistence: the Middle Woodland and Late Woodland transition in the Mid-Ohio Valley. In CM Scarry (ed.): Foraging and Framing in the Eastern Woodlands. Gainesville: University Press of Florida; 138-56

Wymer DA (1994) The social context of early maize in the mid-Ohio Valley. In S Johannessen and CA Hastorf (eds.): Corn and Culture in the Prehistoric New World. Boulder: Westview Press; 411-26.

Yarnell RA (1993) The importance of native crops during the Late Archaic and Woodland periods. In CM Scarry (ed.): Foraging and Framing in the Eastern Woodlands. Gainesville: University Press of Florida; 13-26.

Yarnell RA (1994) Investigations relevant to the native development of plant husbandry in Eastern North America. In W Green (ed.): Agricultural Origins and Development in the Midcontinent Iowa City, IA: Office of the State Archaeologist, Report no. 19 / University of Iowa; pp. 7-24.

The Stressful Revolution: A Rise in Fluctuating Asymmetry from Medieval to Victorian England

Rebecca A. Storm

Biological Anthropology Resource Centre
Department of Archaeological Sciences
University of Bradford
Bradford
West Yorkshire
BD7 1DP
e-mail address for correspondence: rstorm@bradford.ac.uk

Abstract

The repercussions of the Industrial Revolution in England led to the Victorians enjoying a time of prosperity through advancements made in technology, medicine and, for some, living conditions. For all the benefits to society, this was also a devastating time, a time which witnessed a decline in health status and an increase in environmental stresses resulting from pollution, overcrowding, overworking, dangerous working conditions, and poverty. Such adverse living conditions result in growth disturbance (development instability) in biological tissues, including those of the skeleton. Developmental instability can be evaluated though the presence and extent of fluctuating asymmetries that occur within a population. In this study, fluctuating asymmetry was assessed through one hundred and twelve bilateral cranial and post-cranial measurements taken on 456 adults from five English skeletal populations: two Medieval skeletal populations from Chichester, West Sussex and Fishergate, York; a late Medieval and a Victorian site from Hickleton, South Yorkshire; and a Victorian cemetery from Wolverhampton. The results indicate that there was a 6% rise in fluctuating asymmetry levels from the Medieval to the Victorian periods in England. Overall mean asymmetry scores for the Victorian populations were found to be significantly different (t=5.589, p<0.0001) to those of the Medieval populations. The Victorian skeletons reflect increased asymmetry that is especially evident in the scapulae, humeri, radii, femora, tibiae and tarsals. It is the conclusion of this study that along with environmental and social changes wrought by the Industrial Revolution, there was also a reflected increase in developmental instability in human populations.

Keywords: Fluctuating asymmetry; developmental instability; Industrial Revolution; Victorian period; osteometry

Introduction

The eighteenth and nineteenth centuries witnessed the birth of the modern world. The aptly named "Industrial Revolution" was both a transformation and revolution in technology and society. This "revolution" encompassed dramatic changes in industrial and agricultural systems, advancements in medicine, and for some, improvement in their wealth and living environment. Ironically, this was also a time of the deterioration of living conditions and public health, leading to a strain on the social fabric that had significant implications for the working classes (Chadwick 1965; Gaskell 1833; Haley 1978; Porter 1998). Many individuals during this period lived in overcrowded areas with poor sanitation and increased pollution. In 1842 Edwin Chadwick produced an eye-opening report into the lives of the working class. He concluded the report with a disturbing picture:

'...the various forms of epidemic, endemic, and other disease caused, or aggravated, or propagated chiefly amongst the labouring classes by atmospheric impurities produced by decomposing animal and vegetable substances, by damp and filth, and close and overcrowded dwellings prevail amongst the population in every part of the kingdom, whether dwelling in separate houses, in rural villages, in small towns, [or] in the larger towns....the annual loss of life from filth and bad ventilation are greater than the loss from death or wounds in any wars in which the country has been engaged in modern times.' (Chadwick 1965, p. 422).

These adverse living conditions are reflected in growth disturbances to biological tissues, including those of the skeleton. Such developmental instabilities can be evaluated though the presence and extent of fluctuating asymmetries in osseous material that occur within an individual and a population. Fluctuating asymmetry (FA) is variation in the right and left sides of a bilateral structure. These variations are random, independent and usually the differences are small, being less that 1% of the measurable trait within any given population (Møller & Swaddle 2002; Palmer 1994; Palmer & Strobeck 2003; Van Valen 1962). FA occurs when the development of a bilateral structure, which usually develops each side as a mirror image of the other, is disrupted (Larsen 1997). This disruption is caused by a variation in the environment of the developing structure and is usually termed 'biological and/or developmental noise.' Such disturbances have been attributed to pathological processes, genetic predispositions, congenital abnormalities, environmental influences, and/or biomechanical stresses. If postulated that the greater the developmental stability, the greater the symmetry; one could hypothesise that any movement away from symmetry reflects environmental instability and can be measurable (Fields et al. 1995; McManus 1982; Palmer 1994). Moreover, by using calculations of fluctuating asymmetry as measurement of instability, only a sample of that population is needed to produce reliable results (Møller & Swaddle 2002). Since most osteological research involves the analysis of archaeological material or test samples of living populations where the whole population can not be realistically studied, fluctuating asymmetry is thus the best indicator for analysing the influence of the environment on the population's health and well-being.

The aim of the current research is to assess the level of fluctuating asymmetry in individuals from three Medieval and two Victorian populations. The hypothesis is that environmental and social changes during the Industrial Revolution had an adverse affect on individuals' developmental stability and will be reflected through an increase in FA between the two periods. In addition, this study addresses the following questions: 1) What is the normal asymmetry values for the overall sample? 2) Are there any period related differences in the levels of FA between males and females? 3) Are there variations between the social and health status of compared populations?

Materials and Methods

The skeletal samples comprised 456 individual (383 Medieval and 73 Victorian) from five English skeletal populations of differing social, environmental and economic status. Of these individuals, there are 141 females, 307 males, and eight individuals of indeterminate sex. A sample of sixteen individuals came from the cemetery at St. Wilfrid's church, which consisted of a local rural parish population in Hickleton, South Yorkshire, dating from the 12th through to the 19th centuries. Most of the burials were of rectors or parsons of the church and those influential families of the village (Dabell 1999; Sydes 1984). For this study, the sample was divided into two separate populations based on the archaeological phasing of the site: 10 individuals dating from the 17th to 19th century and six from the 12th to 16th century; referred to here as I- Hickleton and M-Hickleton respectively. One hundred and thirty-one individuals were from the Hospital of St. James and St. Mary Magdalene from Chichester, West Sussex. This site was initially founded as a *leprosarium* prior to AD 1118 for a group of eight lepers and eventually converted into an almshouse in AD 1450 as the brethren began accepting women, children, and those inflicted with varying diseases (Lee 2001). Sixty-three individuals came from St. Peter's Church, Wolverhampton, dating from AD 1827 to 1870. Due to the extensive nature of the church's burial records and the city's census, many of the interred families' backgrounds are known. It is known that, when the cemetery was in use, Wolverhampton was a crowded industrial town – with most of the population specialising in the metal trade – and that the economic situation for

the majority of the population was poor (Neilson & Coates 2002). The final sample comprised 256 individuals from two main cemeteries; one dating from the 11th to 12th centuries and the other from the 13th to 16th centuries. Phasing of these two cemeteries indicates that there is a division of burials between the priors of the church, high status families and the general population of lay brethren and parishioners (Stroud & Kemp 1993).

One-hundred and twelve cranial and post-cranial measurements were taken for both the right and left sides, where available (Table 1). Each measurement was taken using either digital sliding callipers, spreading callipers or an osteometric board, where appropriate. Measurements are based on standard cranial osteometric points and previously defined cranial and post-cranial measurements (Howells 1973; Moore-Jansen et al. 1994; Steele & Bramblett 1988; Storm 2000; Storm & Knüsel 2005; White 1991). All measurements were taken by the author to ensure consistency. Asymmetry scores were recorded to the 0.1mm, as differences in asymmetry are usually less than 1% of a trait (Møller & Pomiankowski 1993; Palmer & Strobeck 1986; Palmer 1994; Swaddle et al. 1994). All measurements taken from obviously pathological material were noted and separated from those taken from non-pathological osseous material. Where an element was fragmentary or affected by taphonomical processes, measurements were completed only if they could be taken with confidence. When measurement error was suspected, the measurement was taken at least twice and the mean was recorded. Further, measurement errors of the method were assessed by repeating all measurements ten times on both right and left sides of ten individuals, on ten separate occasions. Differences between the measurements were derived using the formula:

$$TEM = \sqrt{\sum(d^2)/2n}$$

where d is the difference between repeated measurements and n is the number of replicated measurements (Dahlberg 1940; Knapp 1992). Measurement error was found to be insignificant ($p < 0.05$) for all 112 measurements (Table 1).

An absolute value of R-L formula (FA_1) was used to show the degree of any deviation from symmetry within each trait (Palmer 1994; Palmer & Strobeck 1986). Further, a formula correcting for trait size was then calculated (Palmer & Strobeck 1986):

$$FA_2 = |(R-L)| / [(R+L)/2].$$

Once each individual was evaluated, these results were then subjected to the Grubb's test statistic for outliers:

$$T_G = (X_i - \mu)/SD,$$

where X_i is the observed value of the potential outlier, μ is the population mean, and SD is the standard deviation

Table 1. Bilateral measured traits and associated measurement error (TEM=$\sqrt{\sum(d^2)/2n}$).

Measurement	TEM	SD	P
Cranium			
Breadth: orbit (ec-d)	0.22	1.45	0.024
Height: orbit	0.21	2.1	0.01
Cord: n-or	0.32	2.29	0.019
Cord: fmt-n	0.28	2.41	0.014
Cord: fmt-ns	0.22	3.71	0.004
Height: malar	0.26	2.58	0.01
Length: mastoid process	0.3	3.58	0.007
Breadth: mastoid process	0.31	3.74	0.007
Height: mastoid process	0.27	2.47	0.012
Cord: ms-ast	0.19	4.62	0.002
Length: digastric groove	0.31	3.73	0.007
Length: occipital condyle	0.17	2.43	0.005
Cord: ecm-intermaxillary suture	0.24	2.58	0.009
Cord: o-po	0.34	4.08	0.007
Cord: ba-po	0.3	3.57	0.007
Cord: fmt-b	0.28	5.09	0.003
Cord: b-po	0.31	5.15	0.004
Cord: b-zo	0.36	5.52	0.004
Cord: n-ms	0.3	4.64	0.004
Cord: b-ast	0.24	5.72	0.002
Cord: l-fmt	0.34	6.89	0.002
Cord: l-ast	0.24	5.48	0.002
Mandible			
Length: mandible	0.28	6.1	0.002
Length: corpus	0.25	5.45	0.002
Maximum height: ramus	0.29	5.12	0.003
Maximum breath: ramus	0.23	3.49	0.004
Minimum breadth: ramus	0.23	2.75	0.007
Clavicle			
Maximum length	0.23	9.26	0.001
Maximum diameter: midshaft	0.16	1.71	0.009

Measurement	TEM	SD	p
Clavicle cont.			
Minimum diameter: midshaft	0.12	1.22	0.009
Maximum width: acromial end	0.26	3.56	0.006
Maximum width: sternal end	0.16	3.39	0.002
Maximum depth: medial curve	0.41	3.26	0.016
Maximum depth: lateral curve	0.28	2.95	0.009
Scapula			
Length: glenoid fossa	0.34	3.47	0.01
Breadth: glenoid fossa	0.22	2.66	0.007
Maximum length: acromion	0.28	5.41	0.003
Maximum length: coracoid process	0.19	3.51	0.003
Humerus			
Maximum length	0.23	18	<0.001
Maximum diameter: midshaft	0.22	2.03	0.012
Minimum diameter: midshaft	0.14	1.86	0.006
Maximum diameter: deltoid tuberosity	0.3	2.21	0.018
S-1 diameter: head	0.2	3.37	0.003
A-P diameter: head	0.24	3.1	0.006
Breadth: epicondylar	0.08	5.44	<0.001
Anterior breadth: articular surface	0.3	3.64	0.007
Posterior breadth: articular surface	0.29	2.52	0.013
Length: greater tubercle	0.28	2.75	0.01
Radius			
Maximum length	0.3	16.6	<0.001
Maximum diameter: midshaft	0.16	1.62	0.01
Minimum diameter: midshaft	0.16	1.1	0.021
Maximum diameter: head	0.17	1.94	0.008
M-L width: distal end	0.24	2.59	0.008
A-P width: distal end	0.27	2.17	0.016
Ulna			
Maximum Length	0.24	15	<0.001
Physiological length	0.33	14.2	0.001

Table 1. Continued

Measurement	TEM	SD	p
Ulna cont.			
Maximum diameter: midshaft	0.2	1.84	0.012
Minimum diameter: midshaft	0.19	1.38	0.019
Maximum diameter: nutrient foramen	0.26	1.89	0.019
Height: radial notch	0.29	3.53	0.007
Width: olecranon	0.26	2.57	0.01
Height: coronoid	0.28	2.86	0.009
Metacarpals			
MC1: Maximum length	0.07	2.72	0.001
MC2: Maximum length	0.05	4.4	<0.001
MC3: Maximum length	0.13	4.7	0.001
MC4: Maximum length	0.12	4.29	0.001
MC5: Maximum length	0.12	3.49	0.001
Sacrum			
Minimum breadth: ala	0.32	4	0.006
A-P width: ala	0.15	5.73	0.001
Maximum A-P length: auricular surface	0.41	4.97	0.007
Maximum S-I width: auricular surface	0.41	5.73	0.005
Height: body of S1	0.23	4.07	0.003
Os coxae			
Height	0.32	13.6	0.001
Breadth: ilium	0.33	10	0.001
Length: pubis	0.38	5.38	0.005
Length: ischium	0.39	8.38	0.002
Height: acetabulum	0.36	3.72	0.009
Maximum width: symphyseal face	0.3	2.63	0.013
Height: auricular surface	0.38	4.81	0.006
Breadth: auricular surface	0.36	5.88	0.004
Femur			
Maximum Length	0.27	26.1	<0.001
Maximum diameter: midshaft	0.27	2.93	0.009
Minimum diameter: midshaft	0.19	2.14	0.008
	0.34		0.012

Measurement	TEM	SD	p
Femur cont.			
Minimum diameter: subtrochanteric	0.37	2.67	0.019
Breadth: Epicondylar	0.35	5.68	0.004
Maximum length: lateral epicondyle	0.31	4.4	0.005
Maximum A-P diameter: head	0.16	3.57	0.002
Maximum S-I diameter: head	0.2	3.83	0.003
Maximum width: proximal end	0.19	7.48	0.001
Least A-P diameter: neck	0.31	2.89	0.012
Least S-I diameter: neck	0.28	3.57	0.006
Tibia			
Maximum length	0.42	21.7	<0.001
Maximum diameter: nutrient foramen	0.3	3.39	0.008
Minimum diameter: nutrient foramen	0.22	2.56	0.007
Maximum breadth: distal end	0.4	3.93	0.011
Maximum A-P width: distal end	0.42	3.92	0.011
Maximum breadth: proximal end	0.43	5.82	0.005
Maximum A-P width: proximal end	0.39	4.57	0.007
A-P diameter: medial condyle	0.34	3.51	0.009
A-P diameter: lateral condyle	0.3	3.27	0.008
Calcaneus			
Maximum length	0.32	4.91	0.004
Maximum breadth	0.3	3.25	0.008
Maximum height	0.32	3.37	0.009
Talus			
Maximum length	0.4	4.09	0.01
Maximum breadth	0.3	3.16	0.009
Maximum height	0.29	2.26	0.017
Metatarsals			
MT1: Maximum length	0.17	3.58	0.002
MT2: Maximum length	0.11	4.08	0.001
MT3: Maximum length	0.06	3.93	<0.001
MT4: Maximum length	0.1	3.99	0.001
MT5: Maximum length	0.09		

(Palmer & Strobeck 2003). Any significant outliers (P<0.05) were subsequently removed from further analysis as these individuals would skew any results (Palmer & Strobeck 2003). Asymmetries were then tested by Student's t-test for any significant differences between samples.

Results

The results, summarized in Table 2, indicate that all individuals within each of the populations exhibit a degree of asymmetry for each of the measured traits. The overall mean absolute asymmetry score of all 112 measurements for all five populations was found to be 1.0mm and a FA_2 score of 3.3%. Wolverhampton was found to exhibit the most asymmetry with a FA_2 score of 3.8% and a FA_1 score of 1.2mm. The mean fluctuating asymmetry score for the I-Hickleton sample is at 1.2mm, but is 0.4% less than Wolverhampton when asymmetry is corrected for trait size, as its FA_2 score is 3.4%. For the Medieval sites, Chichester has the highest asymmetry of 3.5% with a FA_1 score of 1.1mm, followed by M-Hickleton at 1mm and with a FA_2 score of 2.8%. Fishergate had the lowest overall asymmetry of all the populations with an absolute asymmetry of 1mm and a FA_2 score of 3%.

Comparisons made between the populations indicate that there is an overall difference in asymmetry scores between Medieval and Victorian sites. Student t-test scores indicate there is little comparison (p<0.0001) for both FA_1 and FA_2 between the populations for pooled element asymmetry scores, denoting that the populations are indeed from separate environments. Tables 2-4 and Figure 1 indicates there is a small but significant difference between the two considered populations. These results reveal that of the analysed populations, it is the affluent Medieval site of Fishergate that was found to be as the least asymmetric and Wolverhampton, dating to the Victorian period, the most. For the Medieval samples, mean asymmetry scores for all elements lie close to/or below the overall combined population mean; whereas, the Victorian samples are consistently above the overall mean (Figure 1). Student's t-tests indicate there were significant differences for absolute asymmetry between the two periods for many of the elements: the clavicle, scapula, femur, and tarsals with p<0.05; the cranium, radius, and tibia with p<0.01; and the humerus and overall asymmetry with p<0.001 (Table 3).

Table 2. Population asymmetry scores.

Population	N	FA₁*		FA₂**	
		Mean	SD	Mean	SD
Wolverhampton	63	1.2	0.5	0.038	0.01
Industrial Hickleton	10	1.2	0.2	0.034	0.003
Medieval Hickleton	6	1	0.3	0.035	0.003
Chichester	131	1.1	0.3	0.035	0.007
Fishergate	246	0.9	0.3	0.03	0.007
Victorian	73	1.2	0.5	0.037	0.008
Medieval	383	1	0.3	0.031	0.007
All Populations	456	1	0.3	0.033	0.008

*FA_1= R-L (in mm)
*FA_2=|(R −L)| / [(R+L)/2].

Table 3. Comparison of the FA$_1$ scores for separate time periods (FA$_1$=R-L, in mm).

Element	All Populations			Industrial			Medieval			Medieval v Industrial	
	N	Mean	SD	N	Mean	SD	N	Mean	SD	t-value	p
Cranium	334	1.1	0.7	59	1.4	1	275	1.1	1	3.011	<0.003
Mandible	275	1	0.6	45	1.1	0.7	230	1	0.7	1.21	0.227
Clavicle	308	1.2	0.7	46	1.4	0.8	262	1.1	0.8	2.327	<0.02
Scapula	230	1	0.7	38	1.2	0.9	192	1	0.9	2.179	<0.04
Humerus	335	1.1	0.6	53	1.4	0.7	282	1.1	0.7	3.495	<0.001
Radius	308	0.8	0.5	46	1	0.6	262	0.8	0.6	2.802	<0.006
Ulna	300	1.1	0.6	45	1.1	0.7	255	1.1	0.7	0.45	0.653
Metacarpals	313	0.5	0.2	46	0.6	0.3	267	0.5	0.3	1.585	0.114
Sacrum	238	1.5	0.8	25	1.6	1.1	213	1.5	1.1	0.754	0.452
Os coxae	328	1.3	0.8	39	1.3	0.8	289	1.3	0.8	0.246	0.806
Femur	352	1	0.4	47	1.1	0.4	305	1	0.4	2.077	<0.04
Tibia	312	1	0.5	34	1.2	0.5	278	1	0.5	2.924	<0.004
Tarsals	286	0.8	0.5	29	0.9	0.5	257	0.8	0.5	2.199	<0.03
Metatarsals	271	0.5	0.2	26	0.6	0.7	245	0.5	0.2	1.247	0.214
Mean asymmetry	456	1	0.3	73	1.2	0.5	383	1	0.5	4.828	<0.0001

Table 4. Comparison of the FA$_2$ scores for separate time periods (FA$_2$= |(R –L)| / [(R+L)/2].

Element	All Populations			Industrial			Medieval			Medieval v Industrial	
	N	Mean	SD	N	Mean	SD	N	Mean	SD	t-value	p
Cranium	335	0.035	0.021	58	0.039	0.022	277	0.034	0.02	1.707	0.089
Mandible	335	0.031	0.017	57	0.035	0.017	278	0.031	0.02	1.649	0.1
Clavicle	306	0.061	0.029	44	0.072	0.037	262	0.059	0.03	2.769	<0.01
Scapula	230	0.026	0.016	36	0.03	0.02	194	0.025	0.02	1.662	0.098
Humerus	337	0.03	0.014	54	0.037	0.017	283	0.028	0.01	4.167	<0.0001
Radius	313	0.034	0.017	45	0.037	0.017	268	0.033	0.02	1.591	0.113
Ulna	305	0.035	0.019	48	0.038	0.019	257	0.035	0.02	0.958	0.339
Metacarpals	312	0.04	0.015	45	0.042	0.016	267	0.04	0.01	0.95	0.343
Sacrum	238	0.039	0.025	25	0.042	0.031	213	0.039	0.03	0.582	0.561
Os coxae	327	0.028	0.019	39	0.029	0.021	288	0.028	0.02	0.462	0.645
Femur	355	0.023	0.009	47	0.026	0.009	308	0.023	0.01	2.758	<0.006
Tibia	314	0.022	0.012	34	0.028	0.013	280	0.022	0.01	2.749	<0.006
Tarsals	283	0.016	0.009	29	0.02	0.01	254	0.016	0.01	2.261	<0.02
Metatarsals	360	0.038	0.012	38	0.042	0.01	322	0.038	0.01	2.187	<0.03
Mean asymmetry	456	0.033	0.008	73	0.038	0.01	383	0.032	0.01	5.589	<0.0001

On further analysis of differences between single measurements, the Victorian populations have distinctive asymmetrical changes in the maximum lengths of the upper limb. Maximum lengths of the Victorian humeri were more asymmetrical by 1.6mm; with a Victorian asymmetry score of 5.2mm and Medieval at 3.7mm (t = 2.787, p<0.01). Radial maximum length asymmetry scores were also high with 3.7mm for the Victorians and 2.3mm for Medieval, a 1.4mm difference (t = 3.023, p<0.01). As Figure 2 indicates, the difference in the individual population mean asymmetry can be especially observed in the ulna. This element has one of the highest differences in asymmetry scores, as the physiological maximum length was 4.2mm in Victorians and only 2.9 mm in Medieval (t = 2.681, p<0.01).

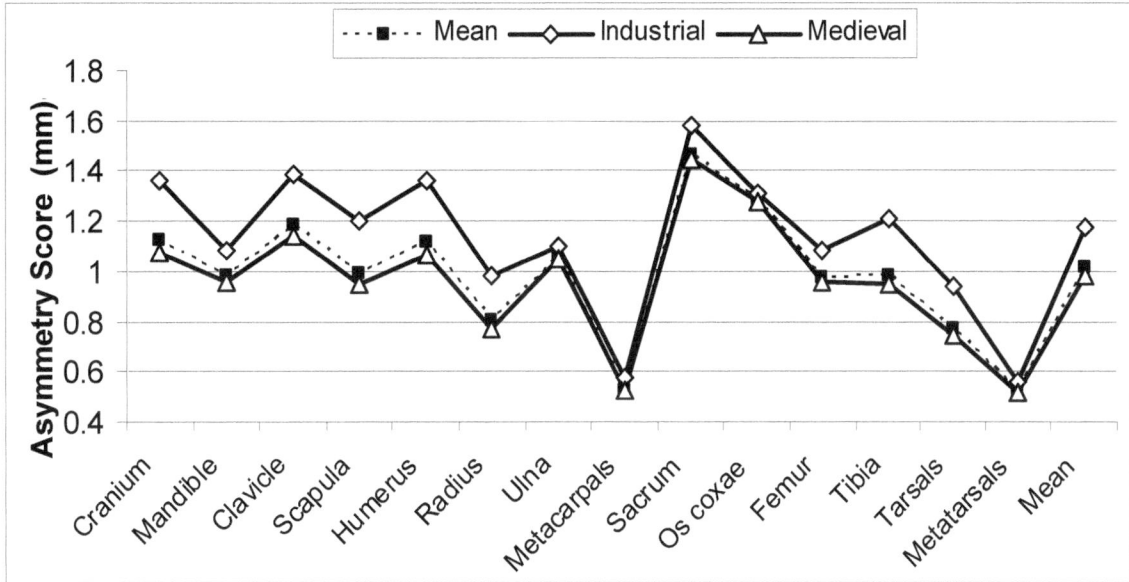

Figure 1. Comparisons of mean asymmetry scores of all the observed populations and the overall mean of R-L

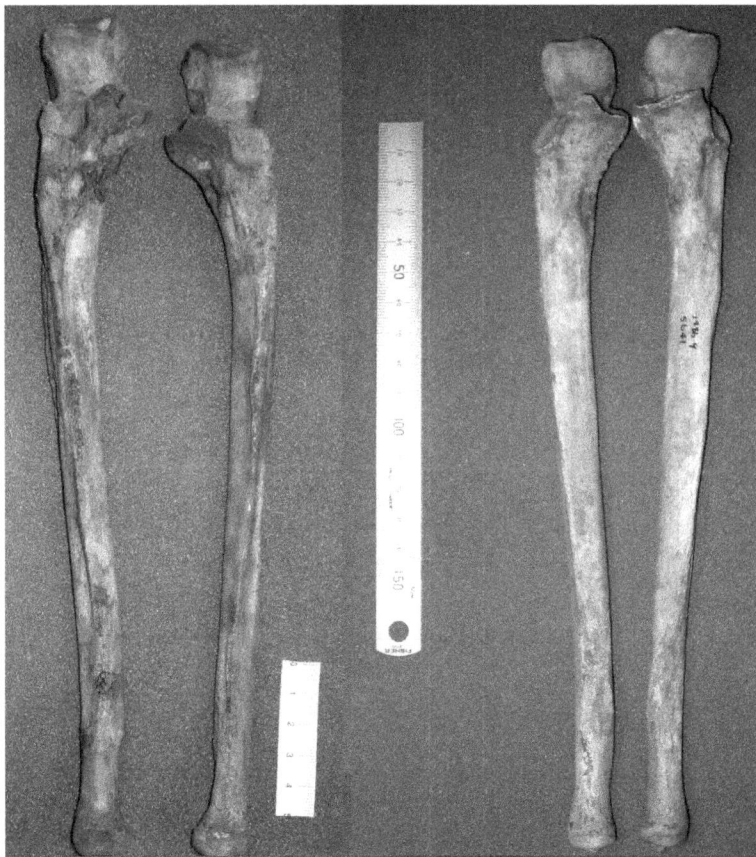

Figure 2. Asymmetry in ulnae from Victorian Wolverhampton (left) compared with Medieval Fishergate (right).
Table 5. Comparisons between males and females over the separate time periods.

T-test Grouping	Valid N	Valid N	Mean	Mean	t-value	p
All females vs. males	141	307	0.034	0.033	1.133	0.258
Industrial females vs. males	37	36	0.038	0.037	0.632	0.529
Medieval females vs. males	104	271	0.032	0.032	-0.249	0.804
Industrial vs. Medieval females	37	104	0.038	0.032	4.202	<0.0001
Industrial vs. Medieval males	36	271	0.037	0.032	3.411	<0.001

Similarly, analysis of FA_2 levels in both males and females indicate a significant difference between the Victorian and Medieval periods (Table 5). Overall, it was observed that there were no significant difference observed between males and females when the two time periods were combined (p=0.258). When individuals from only the Medieval period were assessed, again females and males do not significantly differ (p=0.529). Similarly, no significant differences were found between males and females from solely the Victorian period (p=0.804). However, it was found that females from Victorian populations significantly differed from females of the Medieval populations (p<0.0001), with the same being true for males (p<0.001).

Discussion

The results of this study revealed a significant difference in fluctuating asymmetry scores between all of the studied populations. More significantly, populations from the Victorian period were found to have had higher levels of asymmetry than those from the Medieval period, suggesting that there must have been a catalyst that caused an increase in developmental instability through the progression of time. It has previously been shown in animal models that higher rates of fluctuating asymmetry within a population is in direct relation to developmental instability caused by biomechanical, health, social, and environmental stresses (Fields et al. 1995; McManus 1982; Møller & Swaddle 2002; Palmer 1994). Each of these causes of developmental instability will be discussed in turn.

At least some of the increase in FA observed in this study may be attributable to a higher biomechanical demand on the upper limbs due to changes in labour practices between the two periods. However, the amount of the increase in FA indicates that there was an additional and more influential factor affecting developmental instability found in the Victorian populations. If the increase in FA was due only to differences in biomechanical stress, then it would be expected that there would be a significant difference between the Wolverhampton sample of poor metal workers and the I-Hickleton population that consisted of rectors, parsons and high status individuals. Furthermore, as requirements for stability in the bipedal stance and locomotor functions mitigates against asymmetry in the lower limbs, at the apparent expense of the skull and upper limbs, it would be expected that there would be little changes in FA levels throughout time. However, the results of this

study indicate that individuals dating from the Victorian Period were found to have higher fluctuating asymmetry scores, especially in the upper limb and cranial measurements, than of those from the Medieval period; but there was no significant difference between the two contemporary Victorian populations (t=1.121, p=0.266). Furthermore, there was found to be a significant differences of FA_2 scores (P<0.01), a 3-6% increase, for all lower limb elements between the two time periods. This lower limb asymmetry, and thus the destabilization of the bipedal stance, emphasizes the decrease in developmental stability in the Victorian populations. Therefore, it is concluded that these individuals were under, or affected by, more stress than due to biomechanics alone.

The health status of a population can also be a factor in levels of developmental instability within that population. Those populations with higher amounts of pathologies present could be a reflection of individuals who suffered higher instability during their development. If the increase in FA in the populations were due solely to health status, it would be expected that the Victorian populations would have lower levels of developmental stability than that of the highly pathological population from the Medieval *leprosarium* and almshouse site from Chichester. Although the Chichester population was found to have the highest level of FA for Medieval populations, it was found to be significantly less than the Wolverhampton sample (t=2.640, p<0.009).

The results from this study suggest that social standing of the population may have also had a certain affect on the overall levels of FA. It was found that with a comparison of those individuals from I-Hickleton and those from Wolverhampton, there was only a slight, insignificant, decrease in fluctuating asymmetry found in the more prosperous Hickleton individuals (t=1.121, p=0.266). Unfortunately, with the sample size of the I-Hickleton sample so low, the result can not be confirmed until a greater number of prosperous individuals from the Victorian period can be added to the samples investigated here. Then again, the results indicate that there is a similar slight increase between social strata within the Medieval populations. Although this may be the case, there was a much higher increase in asymmetry between Medieval and Victorian populations than there was between populations from the same time period but of differing social background, suggesting that there was an additional factor in the observed increase in FA.

As documentary evidence suggests that the industrial period is a time in history when the population is known to have been environmentally stressed, and since developmental instability is a result of stress, it is then arguable that the higher levels of fluctuating asymmetry found in the Wolverhampton and I-Hickleton populations were due more to the adverse living environment of the Industrial Revolution than social standing alone. As historians and contemporaries have often mentioned, high status individuals did not escape the poor environmental conditions during the Industrial Revolution. For instance, in 1858 and 1859 the Thames had become so polluted by the lack of proper sanitation that those who worked in the parliamentary offices would have to hang materials soaked in disinfectants to be able to work (Haley 1978). Also, diseases such as cholera and typhus were found to affect all levels of society. Similarly, although high status individuals had better access to foodstuffs, it was found that luxury food items still had been contaminated by bacteria and foreign materials. A report by the Privy Council in 1863 revealed that at least one-fifth of all meat from butchers in England and Wales came from diseased cattle (Bentley 1971; Haley 1978; Porter 1998). As the Wolverhampton sample is made up of individuals from a poverty stricken area, it is suspected that with a future comparison to higher status individuals the results would indicate only slight decrease in, but comparable, fluctuating asymmetry scores similar to the differences observed between the Wolverhampton and I-Hickleton individuals. The indication here is that fluctuating asymmetry levels between the two time periods must have been more affected by environmental changes than by social, biomechanical and health factors. This conclusion is further supported by the lack of significant differences between males and females within each time period, a both male and female fluctuating asymmetry sores were affected at the same rate, increasing with time.

Conclusion

The current study has indicated that there was a significant rise in fluctuating asymmetry from the Medieval to the Victorian period in England. As fluctuating asymmetry has been determined to be a reliable method for the measure of developmental instability, the increased asymmetry found in the Victorian populations can be attributed to documented social and environmental changes wrought by the Industrial Revolution. With this correlation made, there is evidence to support a continuation of the analysis of past societies though period-related asymmetry to establish social, health, and environmental changes throughout human history. Furthermore, as there was found to be significant differences between populations of the same time period, it is also suggested that an inclusion of additional populations would further elucidate the relationship between an individual's social status and asymmetry.

Acknowledgments

I would like to thank Christopher Knüsel and all those at the Biological Anthropology Research Centre (University of Bradford) and Darlene Weston (University of Leipzig) for their help and support. I would also like to thank the editors and anonymous reviewers for their helpful comments and suggestions. Special thanks to the York Archaeological Trust and the University of Birmingham for access to their collection and to the Leakey Trust for its financial support.

Literature Cited

Bentley N (1971) The Victorian Scene: 1837-1901. London: Spring Books.

Chadwick E (1965) Report on the Sanitary Condition of the Labouring Population of Great Britain. 1842. Ed. & Intro. MW Flinn. Edinburgh: Edinburgh University Press.

Dabell JA (1999) St. Wilfrid's Church Hickleton: the Building Development of a Parish Church. Wombell: Thornsby Printers.

Dahlberg G (1940) Statistical Methods for Medical and Biological Students. London: George Allen & Unwin.

Fields SJ, Spiers M, Hershkovitz I, and Livshits G (1995) Reliability of relation coefficients in the estimation of asymmetry. American Journal of Physical Anthropology 96: 83-87.

Gaskell, P (1833) The Manufacturing Population of England, Its Moral, Social, and Physical Conditions, and the Changes which Have Arisen from the Use of Steam Machinery; With an Examination of Infant Labour. London: Baldwin and Cradock.

Haley B (1978) The Healthy Body and Victorian Culture. Cambridge, MA: Harvard University Press.

Howells WW (1973) Cranial Variation in Man: A Study by Multivariate Analysis of Patterns of Differences Among Recent Human Populations. Peabody Museum of Archaeology and Ethnology Papers Vol. 67. Cambridge, MA: Harvard University Press.

Knapp TR (1992) Technical error of measurement: a methodological critique. American Journal of Physical Anthropology 87: 235-236.

Larsen CS (1997) Bioarchaeology: Interpreting Behavior from the Human Skeleton. Cambridge: Cambridge University Press.

Lee F (2001) Hospital of St. James and St. Mary Magdalene, Chichester. Draft Report on the Human Skeletal Remains. Unpublished Report. BARC University of Bradford.

McManus IC (1982) The distribution of skull asymmetry in man. Annals of Human Biology 9: 167-170.

Møller AP and Pomiankowski A (1993) Fluctuating asymmetry and sexual selection. Genetica 89: 267-279.

Møller AP and Swaddle JP (2002) Asymmetry, Developmental Stability, and Evolution. Oxford: Oxford University Press.

Moore-Jansen PM, Ousley SD, and Jantz RL (1994) Data Collection Procedures for Forensic Skeletal Material. Report of Investigations No. 48. Knoxville: University of Tennessee Press.

Neilson C and Coates G (2002) Excavations in Advance of the Extension to the Harrison Learning Centre, University of Wolverhampton, West Midlands 2001: Post-Excavation Assessment and Updated Project Design. Harrison Learning Centre: Wolverhampton.

Palmer AR (1994) Fluctuating asymmetry analyses: A primer. In Markow TA (ed.): Developmental Instability: Its Origins and Evolutionary Implications. Dordrecht, Netherlands: Kluwer; 335-364.

Palmer AR and Strobeck C (2003) Fluctuating asymmetry analyses revisited. In Polak M (ed.): Developmental Instability: Causes and Consequences. Oxford: Oxford University Press; 279-319.

Palmer AR and Strobeck C (1986) Fluctuating asymmetry - measurement, analysis, and patterns. Annual Review of Ecology and Systematics 17: 391-421.

Porter D (1998) The Thames Embankment: Environment, Technology and Society in Victorian London. Akron, OH: University of Akron Press.

Steele SG and Bramblett CA (1988) The Anatomy and Biology of the Human Skeleton. College Station, Texas: Texas A&M University Press.

Storm RA (2000) Metrical Analysis of Asymmetry and Hand Preference in the Pectoral Girdle and Upper Arm as Observed in the Skeleton. MSc Dissertation. Bradford: University of Bradford.

Storm RA and Knüsel CJ (2005) Fluctuating asymmetry: A potential osteological application. In Zakrzewski SR and Clegg M (ed.): Proceedings of the Fifth Annual Conference of the British Association for Biological Anthropology and Osteoarchaeology. BAR International Series 1383. Oxford: BAR Publishing; 113-118

Stroud G and Kemp RL (1993) Cemeteries of St. Andrew, Fishergate. The Archaeology of York: the Medieval Cemeteries 12 (2). Council for British Archaeology: York.

Swaddle JP, Witter MS, and Cuthill IC (1994) The analysis of fluctuating asymmetry. Animal Behaviour 48: 986-989.

Sydes B (1984) The Excavation of St. Wilfrid's Church, Hickleton: A Second Interim Report, September 1984. Sheffield: County Archaeological Service.

Van Valen L (1962) A study of fluctuating asymmetry. Evolution 16: 125-42.

White TD (1991) Human Osteology. San Diego, CA: Academic Press.

Radiography in Palaeopathology: Where next?

Radiography in Palaeopathology

Jo Buckberry* & Sonia O'Connor

Dept of Archaeological Sciences
University of Bradford
Bradford
BD7 1DP
* e-mail address for correspondence: j.buckberry@bradford.ac.uk

Abstract

Radiography has frequently been used during palaeopathological research, and plays an important role in the differential diagnosis of many diseases, including Paget's disease and carcinomas. Traditionally, radiographs were taken in hospitals with clinical equipment. However industrial radiography techniques have gradually become more commonly used, as their superior image quality and improved potential for diagnoses become recognised. The introduction of radiographic scanners has facilitated the digitisation of these images for dissemination and publication. However this is not all that radiographic digitisation can offer the researcher. Digital image processing (DIP) allows the researcher to focus on an area of interest and to adjust the brightness and contrast of the captured image. This allows the investigation of areas of high radio-opacity and radio-lucency, providing detailed images of the internal structures of bone and pathological lesions undetectable by the naked eye. In addition 3D effects, edge enhancement and sharpening algorithms, available through commonly used image processing software, can be very effective in enhancing the visibility of specific features. This paper will reveal how radiographic digitisation and manipulation can enhance radiographic images of palaeopathological lesions and potentially further our understanding of the bony manifestations of disease.

Keywords: radiography, digitisation, digital image processing, palaeopathology, differential diagnosis, curation.

Introduction

Radiography has frequently been used for the diagnosis and interpretation of certain pathologies, for example carcinomas, Paget's disease, fractures and ankylosing spondylitis (Ortner 2003: 503-544 and 574-577; Roberts 2000: 349; Aufderheide & Rodrígues-Martín 1998: 102-3 and 414-5). However radiography has not been used extensively across the whole range of diseases encountered in palaeopathological research, probably due to a combination of cost and access to suitable equipment, both of which can vary from region to region, the emphasis in palaeopathological research still being laid on the macroscopic analysis of skeletal remains alone (Ortner 2002, 6). In contrast, the examination of mummies and bog bodies is more successful in attracting funding and frequently employs the most up-to-date technologies such as Computed Tomography (CT) (Davis 2005: 135-49 and plates 7.1 to 7.5). Clinical radiographs, whilst useful for comparison with archaeological cases, will not present identical images. Soft tissue may mask subtle changes in the bone and in certain conditions superimpose soft tissue lesions over the image of the bone. Consequently the radiographic signature of many bone lesions remains largely unexplored in the archaeological literature. It is also apparent that only a relatively small number of radiographs are published (for example, compare the numbers of photographs and radiographs in any palaeopathological text book: Aufderheide &Rodrígues-Martín 1998; Ortner 2002; Roberts & Manchester 1995), presumably partly due to the difficulties encountered when creating publication standard radiographic images.

The aim of this paper is to show how industrial radiography, applied to the investigation of human bone can enhance our detection, understanding and diagnosis of pathological lesions and why this technique is superior to clinical radiography. The principles of radiographic imaging will be summarised and the advantages of digitising film images for curation, dissemination and image interpretation are discussed and illustrated with a number of palaeopathological case studies.

Industrial radiography is used in the non-destructive evaluation of objects as diverse as metal welds and castings, plastics, fibre reinforced composites and foodstuffs. It is also widely used in the investigation of cultural material including ceramics, paintings, textiles, and archaeological metalwork (Lang & Middleton 2005). Industrial radiography differs from clinical radiography in that it is not governed by the overriding need to protect a patient. Instead the detail of the techniques used in industrial radiography can be adapted to optimise the image quality for each of its varied applications.

Principals of radiographic imaging

X-rays, like visible light, are a form of electro-magnetic energy but they are invisible, travel in straight lines and have much greater energy and shorter wavelengths than light. Consequently, rather than just being reflected or absorbed, they are able to penetrate deeply into, or through, materials that would otherwise be considered opaque. X-rays are also termed 'ionising radiation' because they have the energy to liberate electrons from the atoms of the material through which they are passing, and it is this that can damage living tissues. As a beam of X-rays penetrates an object some of the X-rays will be absorbed or scattered. The amount of this attenuation of the beam will depend on the energy of the X-rays and the atomic number, thickness and density of the material. In a single exposure at a given X-ray beam energy, a thin sheet of lead (a dense material of high atomic number) may appear radio-opaque (white) whilst a dense bone such as a femur (a porous structure of a mixture of elements all with a considerably lower atomic number than lead) will appear more radio-lucent in comparison (shades of grey). Like light, X-rays cause chemical changes in photographic emulsions, related directly to the intensity of the X-ray flux reaching the film and thus can form the image of the object through which they have passed (Halmshaw 1986).

In conventional radiography, also called transmission radiography, the object lies between the source of X-rays and the image receptor – for instance radiographic film. A radiograph is similar to a black and white photographic negative. The emulsion will be darkest where the greatest number of X-rays has reached the film. The more radio-opaque the object, the lighter the emulsion as fewer X-rays will reach the film. It is these variations in the X-ray flux reaching the film that forms the image. This essentially produces an image in two dimensions containing information from the three dimensions of the object which can make the capture of some details and the interpretation of overlapping features in the image problematic. However radiography enables the non-destructive detection and recording of structures hidden within objects.

The ionising effect of X-rays can also be used to record the image without using film. For instance, computed radiography (CR) relies on the use of a photo-stimulatable screen, where electrons moved out of orbit by collision with the X-rays are trapped in a higher orbit, until they are freed by scanning the screen with a laser light. As the electrons fall back to a lower orbit they release energy in the form of light which is detected electronically and converted into a digital X-ray image.

Electrically-powered X-ray units do not produce X-rays of a single energy, but produce a spectrum of energies. For instance a 30 kV exposure is one in which the highest energies of the spectrum do not exceed 30 kV. The lowest energy X-rays, often termed 'soft' or Grenz rays, have a long wavelength and low powers of penetration. Higher energy X-rays, termed 'hard' X-rays, have shorter wavelengths and greater penetrating power.

Industrial versus clinical radiography

With fast films, florescent screens and heavy filtration of the X-ray beam, clinical radiography is designed to minimise the exposure duration and overall X-ray dose to the patient, and to eliminate the lower energy X-rays (below c. 40 kV) which are the most damaging to living tissue. However, this is achieved at the expense of image resolution and contrast.

Industrial radiography is not hampered by the same restrictions. Traditionally the radiography of palaeopathological lesions has been undertaken with clinical equipment but increasingly industrial radiography techniques have become recognised as providing improved potential for diagnoses because of their superior image quality. Beam geometry, energy, and filtration; exposure duration; X-ray dose and film selection in industrial radiography are all focused on producing high definition images. Compared with clinical X-ray images, high quality industrial images will have a much higher resolution, a greater dynamic range and show greater contrast between features of similar radio-opacity. CR produces images with a greater dynamic range than any film, although currently even the highest quality CR screens produce lower resolution images than industrial films such as Agfa D4. However when radiographing archaeological bone in a clinical setting, the use of CR can be preferable over film as it has a greater dynamic range than general medical film and therefore produces higher contrast images.

X-rays can be scattered by interaction with all the materials through which they pass and this scatter causes fogging of the film, which makes the edges of features appear blurred. Very low energy X-rays that cannot penetrate the bone do not contribute to the formation of the image but can form a significant component of the scattered radiation, increasing the fogging. Improvement in the sharpness of image detail can be obtained through careful filtration using thin metal sheet or foil to remove these very low energy x-rays from the spectrum. The improvement in image quality produced by appropriate filtration is illustrated in Figure 1. This shows a first metatarsal with a lytic lesion at the distal metaphysis, which is characteristic of gout. Further lytic lesions were present on the fourth and fifth metatarsals of this individual, from St Mark's Station, Lincoln (medieval Carmelite Friary; Isaac and Roberts 1997). Both radiographs were taken at 60 kV: the right image was taken using a plastic cassette whilst that on the left was taken using an aluminium cassette. The right image has more contrast, but some areas of bone are overexposed and therefore are not recorded on the radiograph (e.g. the medal side of the distal end). In the left image the aluminium cassette has acted as a filter and the detail is clearly sharper. Because the aluminium lid of the cassette is between the object and the film this has not only filtered the primary beam but also removed low-energy scatter generated within the object itself. The filtered image is also slightly lower in contrast but this has improved the rendition of detail as, whilst the detail in the thicker areas of bone is still visible, the thinner areas are not so overexposed.

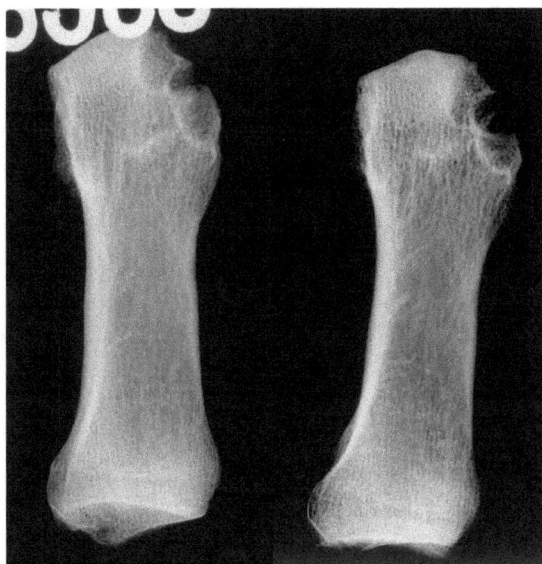

Figure 1. First metatarsal with lytic lesion characteristic of gout (St Mark's Station, Lincoln). The image on the left was taken using an aluminium filter, whereas the image on the right was not filtered.

A step wedge or other agreed standard should be included in all radiographic images to act as an image quality indicator (IQI). This is especially important if bone densities are being investigated. Variations in processing protocols, the condition and temperature of the processing chemicals and the use of different X-ray units can produce variations in the contrast of the images produced, even if the same film and exposure parameters have been used.

O'Connor and O'Connor (2005) discuss the radiography of pathology in bird bones, however many of the observations that they make relating to equipment choice, film selection and radiographic technique are equally valid for the radiography of human remains.

Sharing and publishing images – some solutions

As radiographs became more routine in osteological research, their dissemination has become a priority. Viewing and handling of film radiographs inevitably leads to the degradation of the radiograph. Prolonged exposure to light causes photodegradation of the image and the delicate emulsion is also prone to physical damage, for example scratching the surface and sticky fingerprints, every time the film is slid out of its protective envelope and put on a light box. The reproduction of X-ray images by photographic techniques or by digitising them with a light box and flat bed scanner often produces disappointing results. The introduction of industrial radiographic scanners is revolutionising our ability to capture information in all the tones of the X-ray image in a digital form (jpg, tiff, etc.). Once digitised, the original X-ray image can be archived in suitable controlled storage conditions to ensure its long-term survival. The digital copies can be stored electronically and used for teaching, discussion

with colleagues, printed to film or paper, or reduced in resolution (as appropriate) for inclusion on web sites, publication, posters and museum displays. Through digitisation, endless numbers of identical copies may be produced and can be distributed among members of a research team, for example osteologist, archaeologist and contractor and viewed countless times without degradation of the image quality or loss of information.

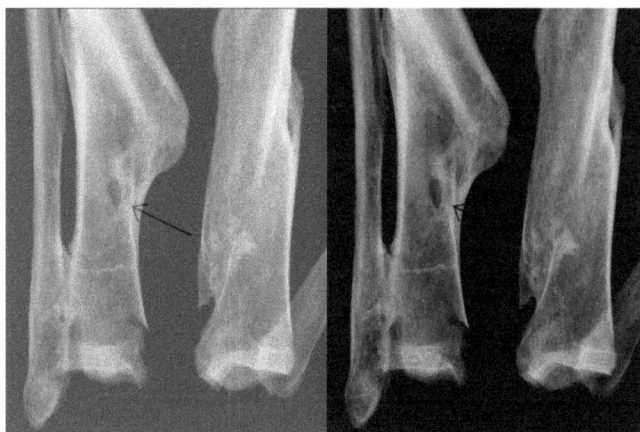

Figure 2. Fractured tibiae and fibulae with osteomyelitis (Baldock, Roman). The old radiograph (left) appears to be deteriorating. Digitisation has preserved the image for our archives, and allows it to be investigated using digital image manipulation.

Figure 2 is of the fractured lower legs of a Roman skeleton from Baldock, Hertfordshire (Roberts 1984). The old radiograph, on the left, has deteriorated over time, probably due to poor chemical processing in combination with extended exposure to light whilst in use as teaching material. The level of contrast, in particular, is much reduced. Digitisation of the radiographic film has preserved the image for our archives, and has allowed it to be investigated using digital image manipulation (right), greatly improving the sharpness and contrast. The arrow in this image pointing towards a cloaca was an annotation made on the film in pen for teaching purposes. Digitisation has removed the need to annotate films directly as the arrow and any additional notes would be made to a digital copy.

Digital Image Processing (DIP)

It is possible to digitise an X-ray film by photographing it on a light box using a digital camera or by scanning it using a conventional flat bed scanner converted to digitise transparent media. However, the dynamic range of X-ray images is very much wider than, for instance, a photographic negative and it is unlikely that either of these approaches will capture detail equally well from all the densities (from black through to white) which have been recorded on the film (O'Connor and Maher 2001). An industrial X-ray film scanner is now used routinely at the University of Bradford.[1] It is designed specifically to

---

[1] Agfa FS50B industrial X-ray film scanner with Radview capture software.

be used to capture the detail in all the optical densities of the film, including variations not distinguishable by the naked eye. Digital image processing (DIP) using commonly available image processing software (such as Adobe Photoshop and Paint Shop Pro) of the captured data allows us to improve the visibility of these features to an extent that was not previously achievable.

DIP is often regarded as a means of falsifying information. This is possible in unscrupulous hands, just as convincing photomontages could always be created by talented photographers. Used responsibly, DIP is a means of improving the visibility of captured information. DIP can be used to rescue information from technically poor X-ray images or those that have degraded over time (Fig. 2), however it cannot be used to create information that was not captured in the first place, and should not be seen as a substitute for good radiographic practice. There is no problem in using digitally manipulated images for comparative purposes where, for instance, only the brightness or contrast has been adjusted to improve feature visibility when there is a suitable IQI in the image. However, if techniques such as sharpening, edge enhancement or texturing have been applied it would be essential that the original image be published alongside the enhanced version.

Magnification of detail or converting the image from positive to negative form are achievable at the click of a button and can reveal details which might otherwise have been overlooked.

Figure 3 is a radiograph of a possible case of osteomalacia from the medieval Blackfriars cemetery in Gloucester (Wiggins et al. 1993). Macroscopically the ossa coxae are paper-thin and very dysplastic, with thumb sized depressions in the centre of the ilia. The rest of the skeleton is notably very light, with very thin bone cortex, and a 'pinched' appearance to many muscle attachment sites, however no micro-fractures have been observed. Digitisation of the radiograph has allowed the researcher to enlarge an area of interest, in this case where the ilium becomes paper-thin. This area displays a total loss of trabecular bone, a sclerotic margin and some ovoid areas of radio-lucency. Could these areas be typical of osteomalacia?

The investigation of very dark or light areas of an image and also the discernment of detail within features which, to the naked eye, may seem to be just an undifferentiated grey can be achieved through reassigning the greyscale values of the pixels which make up the image. This can be done through adjusting the picture's contrast or brightness, or through more complex histogram adjustments. These adjustments are particularly useful in revealing the details of the internal structures of bone and pathological lesions. 3D effects, edge enhancement, sharpening algorithms and pseudo-colour imaging of specific grey levels can also prove effective in highlighting specific features (O'Connor et al. 2002).

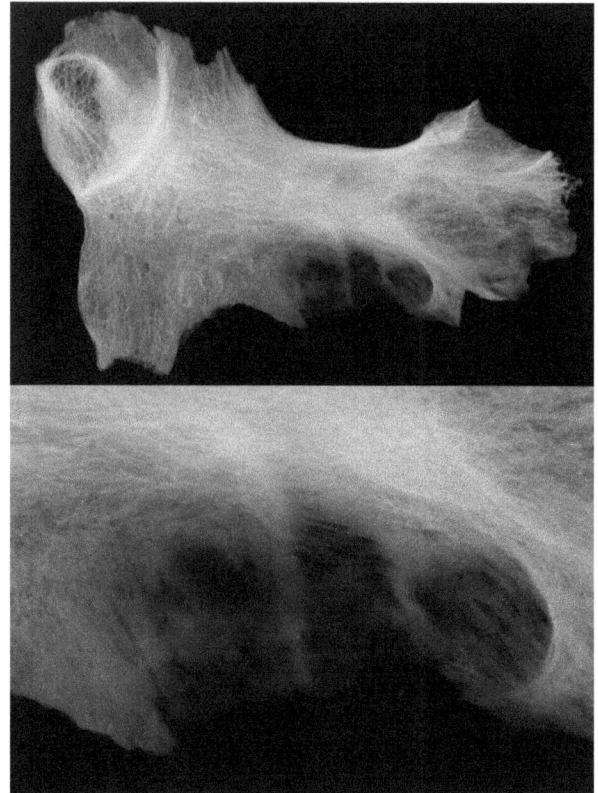

Figure 3. Possible case of osteomalacia from Blackfriars, Gloucester. Digitisation allows the researcher to magnify an area of interest, revealing a total loss of trabecular bone, a sclerotic margin and some ovoid areas of radio-lucency.

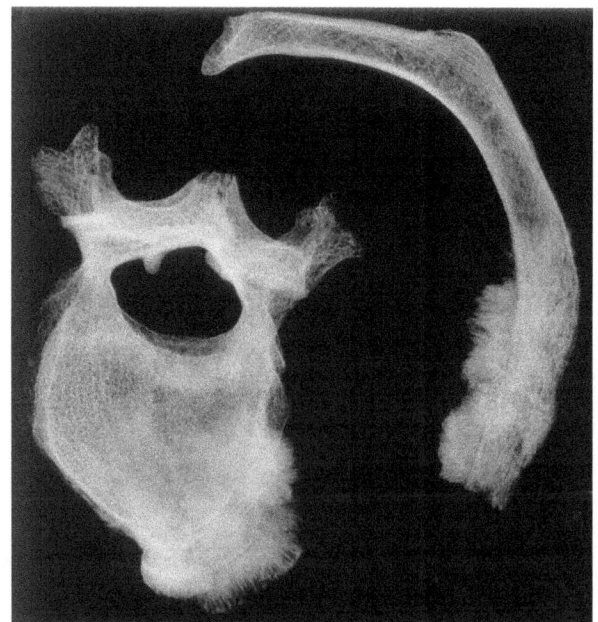

Figure 4. Sunburst lesions on vertebra and rib from St Peter's Wolverhampton.

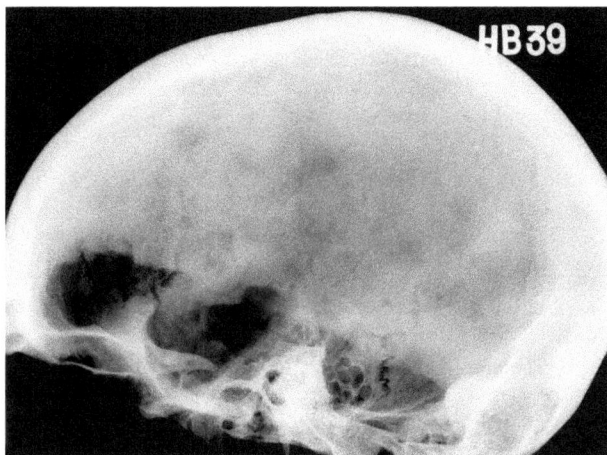

Figure 5. Cranium of individuals from St Peter's Wolverhampton revealing multiple lytic foci.

Figures 4 and 5 are radiographs of the cranium, vertebra and ribs of a 19th-century individual from St Peter's Wolverhampton (Arabaolaza et al. 2005; Adams J and Colls K in prep). Macroscopic analysis of the skeleton revealed the sunburst lesions on the vertebra and ribs (Fig. 4), and multiple small lytic lesions with periosteal new bone formation characteristic of metastases on many bones throughout the postcranial skeleton. A small area of porosity visible on the cranium led to the second radiograph (Fig. 5), which reveals multiple lytic lesions across the vault, many of which are contained within the diplöic space. This is a case of metastatic neoplasm, probably osteosarcoma.

Conclusion

Radiography can reveal pathological lesions even when there is little or no external macroscopic manifestation and provides us with more information about the internal architecture and spread of specific conditions than visual inspection allows. Whilst the radiographic signature of particular diseases like Paget's disease are well understood; others, such as dental granulomas and cysts, are yet to be fully characterised in palaeopathology. Digitising film radiographs using a dedicated X-ray scanner improves their diagnostic potential through the application of digital image processing to enhance the visibility of features of interest. It also facilitates the archiving and dissemination of radiographs and aids the preparation of quality images for publication. Radiography should be used and published more routinely in palaeopathological research.

The palaeopathological community should concentrate less exclusively on external appearances and give much more attention to the internal structures revealed by radiographic images. We should embrace and develop the use of industrial and digital radiographic techniques, free from the anxiety about tissue damage that has bedevilled clinical radiography. By this means a new standard of radiographical excellence may be achieved.

Acknowledgements

We are very grateful to Alan Ogden for his incisive comments on this manuscript and for sharing his insights of the shortcomings of clinical and palaeopathological radiography. We also thank Anthea Boylston, Darlene Weston, Alan Ogden, Chris Knüsel and Holger Schutkowski for their helpful comments on the interpretation of the radiographs, and the two anonymous reviewers for their comments on this paper. Any errors are solely the responsibility of the authors. Anthea Boylston, Iraia Arabaolaza, Paola Ponce and Jo Adams allowed us to use radiographs of material from St Peter's Wolverhampton in advance of publication.

Literature Cited

Adams J and Colls K (eds.) (in prep) 'Out of darkness, cometh light' Excavations in the overflow burial ground of St. Peter's Church.

Arabaolaza I, Ponce P and Boylston A (2005) St Peter's Collegiate Church, Wolverhampton. Report on the human skeletal remains. Bradford: Biological Anthropology Research Centre, University of Bradford; Unpublished.

Aufderheide AC and Rodrígues-Martín C (1998) The Cambridge encyclopedia of human paleopathology. Cambridge: Cambridge University Press.

Davis R (2005) Radiography: archaeo-human and animal remains. Part I: Clinical radiography and archaeo-human remains. In J Lang and A Middleton (eds.): Radiography of cultural material. Oxford: Butterworth Heinemann: 130-149.

Halmshaw R (1986) Industrial Radiography. Mortsel: Agfa-Gevaert.

Isaac L and Roberts C (1997) Cemetery at former St Mark's railway station, Lincoln. Bradford: Calvin Wells Laboratory, University of Bradford; Unpublished.

Lang J and Middleton A (eds.) (2005): Radiography of cultural material. Oxford: Butterworth Heinemann.

O'Connor S and Maher J (2001) The digitisation of X-radiographs for dissemination, archiving and improved image interpretation. The Conservator 25: 3-15.

O'Connor S, Maher J and Janaway R (2002) Towards a replacement for xeroradiography. The Conservator 26: 100-113.

O'Connor T and O'Connor S (2005) Digitising and Image-Processing Radiographs to Enhance Interpretation in Avian Palaeopathology. In G Grupe and J Peters (eds.): Documenta Archaeobiologiae 3: Feathers, Grit and Symbolism. Birds and Humans in the Ancient Old and New Worlds. Rahden: Verlag Marie Leidorf GmbH; 69-82.

Ortner D (2002) Palaeopathology in the twenty-first century. In K Dobney and T O'Connor (eds.): Bones and the man. Studies in honour of Don Brothwell. Oxford: Oxbow; 5-13.

Ortner D (2003) Identification of pathological conditions in human skeletal remains. London: Academic Press.

Roberts C (1984) The human skeletal report from Roman Baldock, Hertfordshire. Bradford: Calvin Wells Laboratory, University of Bradford; Unpublished.

Roberts C (2000) Trauma in biocultural perspective: past, present and future work in Britain. In M Cox and S Mays (eds.): Human osteology in archaeology and forensic science. London: Greenwich Medical; 337-356.

Roberts C and Manchester K (1995) The archaeology of disease. Stroud: Sutton.

Wiggins R, Boylston A and Roberts C (1993) Report on the human skeletal remains from Blackfriars, Gloucester (19/91). Bradford: Calvin Wells Laboratory, University of Bradford; Unpublished.

www.ingramcontent.com/pod-product-compliance
Lightning Source LLC
Chambersburg PA
CBHW061007030426
42334CB00033B/3392

* 9 7 8 1 4 0 7 3 0 1 5 6 3 *